FLEET TACTICS

UNDER

STEAM.

BY

FOXHALL A. PARKER,
CAPTAIN U. S. NAVY.

WITH ILLUSTRATIONS.

BY AUTHORITY OF THE NAVY DEPARTMENT.

NEW YORK:
D. VAN NOSTRAND, PUBLISHER,
23 MURRAY AND 27 WARREN STREET.
1870.

PREFACE.

U. S. Navy Yard,
Boston, *Sept.* 22*d*, 1869.

In his preface to "Squadron Tactics under Steam," published early in 1864, the author remarked: "The 'Naval Warfare' of Sir Howard Douglas, and the 'Tactique Navale' of the French, are, at present, the only works on *steam* tactics deserving consideration." To these must now be added, "Nouvelles Bases de Tactique Navale," by Rear-Admiral Grégoire Boutakov, of the Russian Navy, a book of far greater scientific merit than anything that has preceded it. Believing, however, that a plain, practical work on this subject is still needed, the author has been encouraged to publish this volume, which is simply an adaptation of military to naval tactics, *as put to the practical test* during the two years that he commanded the Potomac flotilla.

"The tactician," says Ramatuelle, "is one who is gifted with quickness of apprehension and clearness of thought, and such correctness of judgment as shall enable him to make choice of the movements best suited to the time and situation." In other words, *great* commanders, by sea as well as on land, are "born, not made;" yet an excellent "recipe for a *good* admiral" is that given by Rear-Admiral Charles Ekins, "in the person who combines theory with practice, is blessed with a clear head, and has *his heart in the right place.*"

"In case signals cannot be seen or clearly understood, no captain can do very wrong if he places his ship alongside that of an enemy."—NELSON.

FLEET TACTICS UNDER STEAM.

An assembly of twelve or more vessels takes the name of fleet, and is separated into three divisions, of one, two, or three squadrons each; each squadron comprising not less than four vessels. Thus, a fleet of twelve vessels would be composed of three divisions of one squadron each, and the commanding officers of squadrons would also be the commanding officers of divisions ; and so with a fleet of any number of vessels up to twenty-four. In a fleet of this or larger size, the divisions would consist of two or more squadrons each, with the commands thus distributed.

Commander-in-Chief. . . . The Fleet.
2d in command. . . . . Van Division.
3d " " . . . . Rear "
4th " " . . . . Centre "
5th " " . . . . Van Squadron.
6th " " . . . . Rear "

7th in command	....	Right Centre Squadron.
8th " "	....	Left " "
9th " "	....	Next to Van "
10th " "	....	Next to Rear "
11th " "	....	2d from Van "
12th " "	....	2d " Rear "
13th " "	....	3d " Van "
14th " "	....	3d " Rear "

And so on, *ad infinitum.*

When the centre division has in it an uneven number of squadrons, there will be of course but one centre squadron, commanded by the 7th in command. The 8th in command then taking the next to van squadron. The 9th the next to rear, etc., etc., etc.

The commander-in-chief of a fleet and the commanding officers of divisions, consisting of *two or more squadrons each,* should have no fixed position, either in line or in column, but be at liberty to move about, from point to point, as the exigencies of service or battle may require. They should ordinarily be found, however, near the centre of their commands.

Commanders of squadrons are on the right of their squadrons, in line, and in the advance, in column, in a fleet in natural order, with the exception of the commander of the rear squadron, who is on the left, in line, and in column brings up the rear, so that in order reversed, that is with the *left* (or rear) *in front*, he leads.

In time of war, each division should be accompanied by a few light-armed vessels of great speed, to be employed as look-out and despatch vessels; and two such vessels should always accompany the vessel bearing the flag of the commander-in-chief.

In order of battle, the reserve-division should, as a general rule, be about one-fourth of the strength of the whole fleet, and be composed of vessels taken equally from the van, rear, and centre divisions. Then, upon a signal being made to the reserve to reinforce *the fleet*, the vessels of it repair to their respective divisions, and when a particular squadron or division is to be strengthened, the commander-in-chief will signal accordingly.

There are but three formations for a fleet, any one of which, according to circumstances, and the vessels of which it is composed, may constitute an order of battle, viz.:—

Line (Fig. 1), Column (Fig. 2), Echelon (Fig. 3).

Fig. 1.

The order of battle for *iron-clads, rams, and torpedo vessels:*

Fig. 2.

The order of battle for vessels whose fighting power is in their *broadside batteries:*

Fig. 3.

Orders offensive and defensive for *vessels of all descriptions:*

Vessels are said to be in *direct echelon* when, steering the same course, each bears from its next astern,* at an angle of 45° (4 points) from the course; consequently, the wings of a fleet, in double echelon, form a right angle (Fig. 3); and this is always to be understood as the bearing upon the signal: *Form echelon* or *double echelon*, unless the commander-in-chief signals the bearing.

By moving a number of vessels (1, 2, 3) in line (Fig. 4) or in column (Fig. 5), through the arcs of circles of equal radii, or upon their centres as axes, it will be observed that, when steering a course at right angles to their original one, they are in column *if moved from line*, or in line *if moved from column*, while at sixteen points they resume their original formation (reversed), and at 4, 12, 20, and 28 points are in echelon. In all these formations the *line of bearing* remains unaltered.

Fig. 4.

Fig. 4 represents the vessels (1, 2, 3) *in line* at the *commencement* of their manœuvres:

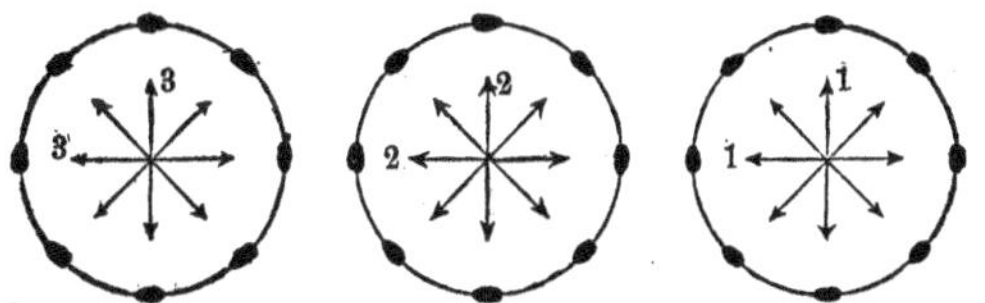

* The *reciprocal* bearing of two vessels in *direct echelon* will of course be 45° and 135° from the course.

FIG. 5,

representing the three vessels (1, 2, 3) *in column*, at the *commencement* of their manœuvres :

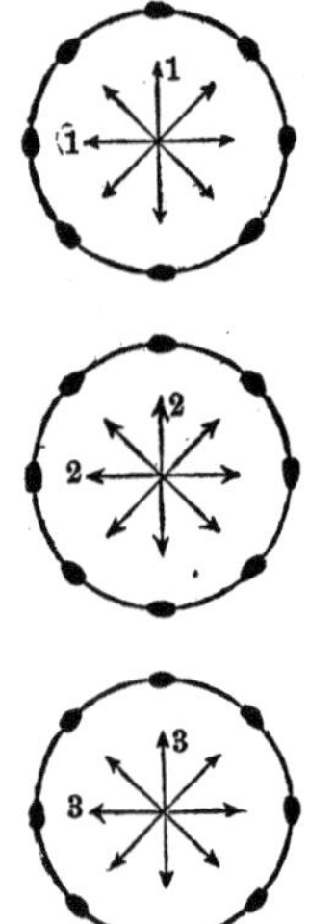

Close order for vessels is one cable's length, or one hundred and twenty fathoms, *from mainmast to mainmast;* open order is two cables' length ; half distance is sixty fathoms. The distances of vessels from each other, in every formation, are as follows :

TABLE A.

FORMATION.	HALF DISTANCE.		CLOSE ORDER.		OPEN ORDER.	
	Laterally.	Longitudinally.	Laterally.	Longitudinally.	Laterally.	Longitudinally.
Line.........	½ cable	....	1 cable	....	2 cables	....
Column of vessels	....	½ cable..	....	1 cable ..	....	2 cables.
Double column ..	½ cable	1 cable...	1 cable	2 cables..	2 cables	4 "
Triple columns...	"	1½ cables	"	3 " ..	"	6 "
Column of fours...	'	2 cables..	"	4 " ..	"	8 "

and so on, *ad infinitum*.

By moving a number of vessels, *in any order*, from line into column to the right or left, the above will be made apparent. Figs. 6, 7, 8, and 9.

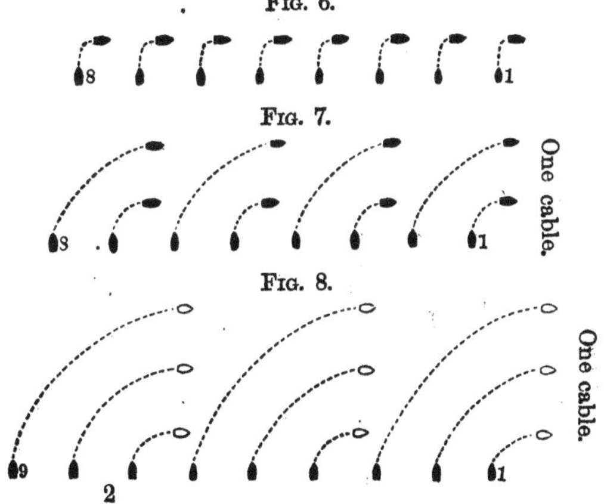

FIG. 6.

FIG. 7.

FIG. 8.

FIG. 9.

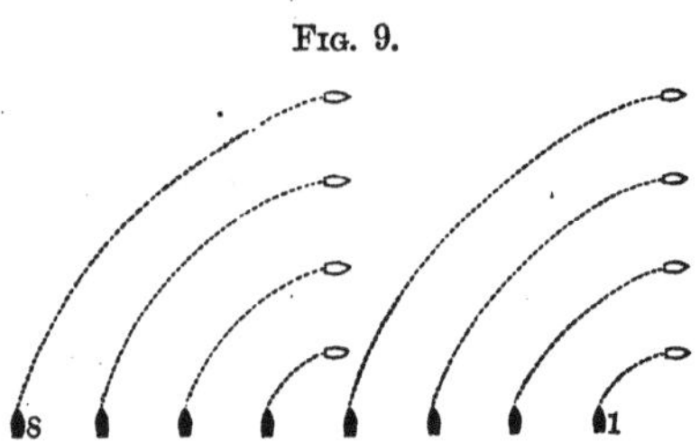

The distances from each other of the divisions or squadrons of a fleet, steaming in order abreast, depend on their formation, as shown by Figs. 19, 23, 27, etc.; the object being, in all formations, to enable the fleet to deploy into line to the front, rear, right, or left, with its vessels in their proper positions. Figs. 100, 102, 105, 106, 108, 109, etc.

The commander-in-chief may signal the fleet to *close up* or *close in*, however, as his judgment dictates.

When vessels are thrown into echelon from line or column, by turning to the right or left, 4, 12, 20, or 28 points, their distances remain the same, provided they have described equal arcs. (Figs. 4 and 5.) When they move forward into echelon, however, upon the signal, *Form "echelon!"* (as in Figs. 10 and 11) their distances are in close order 170, in open order 340 fathoms, and at half distance 85 fathoms.*

Fig. 10—*Close order.*

Fig. 11—*Open order.*

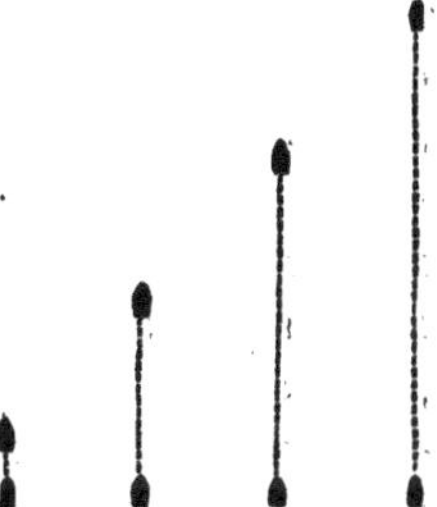

A fleet is in natural order with the van on the right, or leading, and in order reversed when the contrary is the case.

One vessel should always be designated by sig-

* Mathematically stated, these distances are :
169.9+ 339.4+ 84.8+

nal to act as a guide, by which the movements of the other vessels are to be governed, and should wear a guide flag at the main. This, of course, applies only while the fleet is *en route;* when it is manœuvring, the vessels upon which a formation is made must necessarily be the guides.

The rate of speed of a fleet *en route* must be regulated by signal from the commander-in-chief; *steerage-way* is four knots an hour.

The commander-in-chief of a fleet may arrange and number the vessels thereof, as in his judgment may seem best.

In the following problems I have supposed the fleet to be in natural order, to be numbered from right to left, and to be heading North, simply for convenience in demonstrating.

Figs. 12, 13, 14, and 15 show the order of a fleet of twenty-four vessels, formed into three divisions of two squadrons each, in line, column, and echelon.

FIG. 12.

Fleet in line, in natural order, that is, with the van squadron on the right.

This order is reversed by each vessel turning sixteen points to the right or left, so that the rear squadron takes the right.

Rear Squadron.	Next to Rear.	Left Centre.	Right Centre.	Next to Van.	Van Squadron.
Rear Division.		Centre Division.		Van Division.	

FIG. 13.

Fleet in column, in natural order, that is with the van squadron leading. This order is reversed by each vessel turning sixteen points, to starboard or port, so that the rear squadron leads.

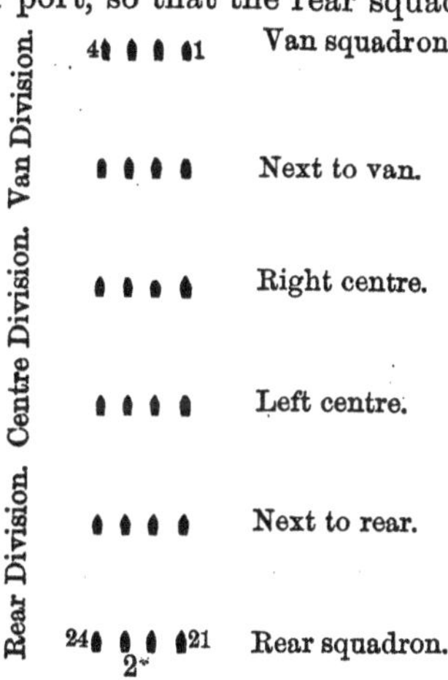

Figs. 14 and 15.

Fleet in double-echelon, in natural order, that is with the van squadron on the right. This order is reversed by each vessel turning sixteen points to starboard or port, so that the rear squadron is on the right.

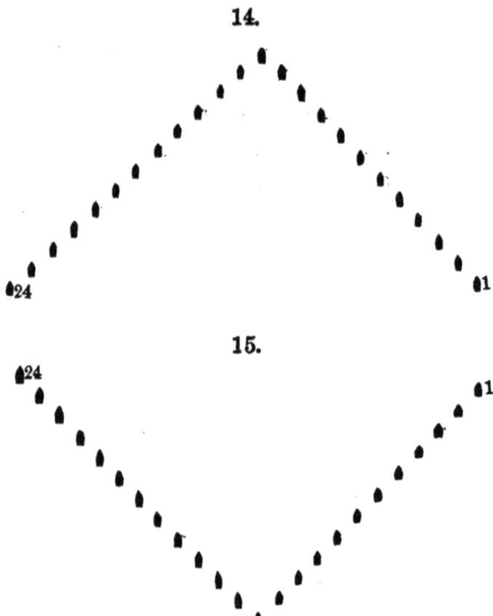

1. The fleet being in line, heading North, to form it into column of vessels, from the right, preserving the original direction.

1*st Method.*

FIG. 16.

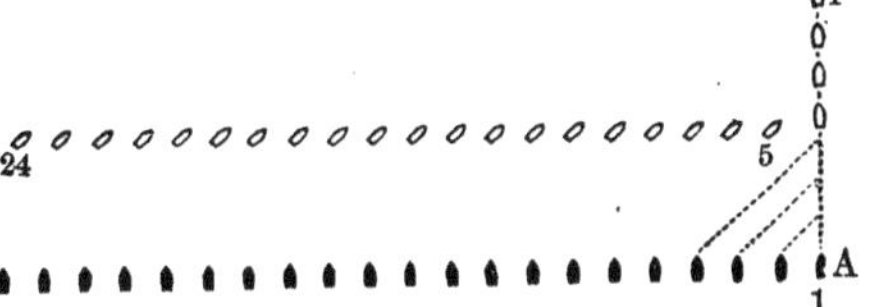

Commander-in-chief signals:

*From the right of fleet—form column of vessels—fleet N.E.—right vessel N.**

Flag-ship of van division: *Division N.E.*†—*right vessel** *N.*

Flag-ships of centre and rear divisions signal: *Division N.E.*†

Upon the hauling down of these signals, the vessel on the right of the fleet keeps its course; the other vessels steer N.E.; two resuming its original direction when in the wake of one, three

---

* Right vessel's distinguishing pennant must here be displayed *over compass signal N.*

† Speed signals should always be made by the divisional commanders to their divisions.

when in the wake of two, and so on to the last vessel.

2d Method.

Fig. 17.

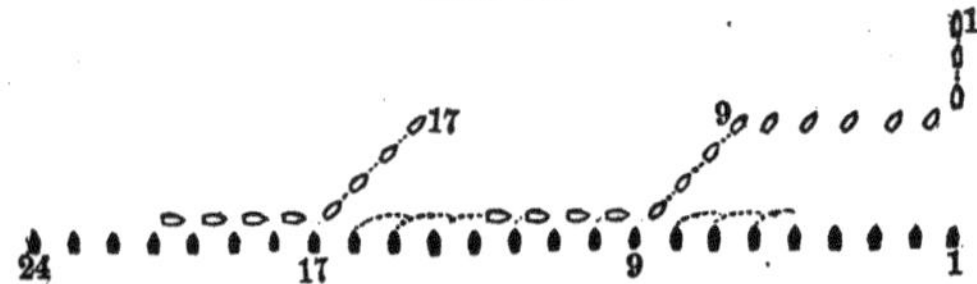

The commander-in-chief signals :

From the right of fleet, form column of vessels—fleet N.E.—right vessel N.

Flag-ship of van division : *Division N.E.—right vessel N.* ; and comes into column as in the 1st method.

Flag-ships of centre and rear divisions signal: *Division E.—right vessel N.E.*

Upon the hauling down of these signals, the vessel on the right of the fleet keeps its course; the vessels on the right of the centre and rear divisions steer N.E., the other vessels E., those of the van division resuming their original direction when in the wake of one, while those of the centre and rear divisions steer N.E. when in the wake of nine and seventeen, respectively, and come into column as before.

3d Method.

FIG. 18.

Commander-in-chief signals:

From the right of fleet, form column of vessels—fleet E.—right vessel N.*

Flag-ship of van division: *Division E—right vessel N.*

Flag-ships of centre and rear divisions signal: *Division E.*

Upon the hauling down of these signals, the vessel on the right of the fleet keeps its course; the other vessels steer E.; two resuming its original direction, when in the wake of one; three when in the wake of two, and so on to the last vessel.

It is evident that the fleet can be formed into column, from the left, preserving the original direction, according to the same principles.

When the commander-in-chief desires to leave it optional with the divisional commanders to

* Speed signals should always be made by the divisional commanders to their divisions.

come into column either by obliquing or taking the two sides of the triangle, he omits signalling the course to the fleet thus:

*From the right of fleet, form column of vessels—right vessel N.**

NOTE TO 1ST METHOD (FIG. 16).—In close order, the distance between the right and left vessels of a fleet of twenty-four vessels, in line, is 2,760 fathoms; it follows, then, that the left vessel steering N.E., and thus making an angle of 45° with the course, will traverse 3,903 fathoms and strike the perpendicular to the line described by the right vessel, moving N., at a distance of 2,760 fathoms from A, the point of departure of that vessel when, if both vessels have proceeded at the same rate of speed, the distance between the van and the rear will be but 1,143 fathoms, whereas it should be 2,760 fathoms (see Table A), to insure which distance the right vessel should have made 5,520 fathoms. Now 5,520 : 3,903 :: 10 : 7.07—therefore the speed of the left vessel should be but $\frac{7}{10}$ths of that of the right. Again, the second vessel, 120 fathoms from the first, will traverse 169.7 fathoms, and strike the perpendicular at a distance of 120 fathoms from A, when the first vessel should be 120 fathoms ahead of her, and have made 240 fathoms. But 240 : 169.7 :: 10 : 7.07—therefore, the speed of the second vessel must also be $\frac{7}{10}$ths of that of the right vessel; *and so with the other vessels, each one resuming her speed when she comes N., in the wake of the leading or guide vessel.*

This seems self-evident, and yet an eminent French tac-

* This paragraph applies to the succeeding problems also.

tician, in forming column in this way, directs the leader of the column to "slow" thus :

"Lorsque l'evolution est dessinée, la tete de la ligne diminue de vitesse pour accélérer le mouvement des vaisseaux de queue."—("Tactiqe supplémentaire a l'usage d'une Flotte Cuirassée." By Vice-Amiral Cte. BOUËT-WILLAUMEZ, Commandant-en-Chef l'escadre d'evolutions.)—Evolution No. 5.

The annexed table shows the rate of speed for each ½ point of obliquity from the course, supposing the speed of the leader (or leaders, since it is evident that the same reasoning applies whatever may be the *front* of the column; a double column, for instance, being simply *two* "columns;" a triple column being *three* "columns of vessels," etc.) of the column to be ten knots an hour, from which every captain—the speed of the leader of the column being known—can readily deduce that of his own vessel. For instance, let the line, heading N. and steaming 15 knots an hour, be ordered to form column, from the right, preserving the original direction, and the course signalled by the divisional commanders to the obliquing vessels be N.N.E. Then 10 : 15 : : 7.6 : 11.4; and we have for the obliquing vessels 11 knots and $\frac{4}{10}$ an hour.

TABLE B.

Angle of Obliquity in Degrees and Parts of a Degree.	Angle of Obliquity in Points and Half points.	Speed of Obliquing Vessels in Knots and Tenths.	Speed of Leading or Guide Vessel (or Vessels) in Knots.
5° 37′ 30″	½ point N. ½ E.......	9.2	10
11 15 00	1 point N. by E.........	8.5	10
16 52 30	1½ points N. by E. ½ E..	8.0	10
22 30 00	2 points N. N. E........	7.6	10
28 07 30	2½ points N. N. E. ½ E...	7.4	10
33 45 00	3 points N. E. by N.	7.2	10
39 22 30	3½ points N. E. ½ N......	7.1	10
45 00 00	4 points N. E.	7.0	10
50 37 30	4½ points N. E. ½ E.	7.1	10
56 15 00	5 points N. E. by E......	7.2	10
61 52 30	5½ point N. E. by E. ½ E.	7.4	10
67 30 00	6 points E. N. E........	7.6	10
73 07 30	6½ points E. N. E. ½ E...	8.0	10
78 45 00	7 points. E. by N........	8.5	10
84 22 30	7½ points E. ½ N....	9.2	10

When the commander-in-chief desires to form the column in the shortest possible time, without regard to the positions of the vessels in it, he has but to make general signal *full speed.* It will sometimes happen, however, that vessels coming into column thus will interfere with each other, when the junior officer must *slow down* until his superior has passed him.

With a fleet of more than twelve vessels, the angle formed by the course of the obliquing vessels with that of the leader of the column, should be at least two points, since if it be less the fleet will be too long in forming. For instance, twenty-four vessels in line, in close order*, heading N., will be three hours in forming 'column of vessels," steering N. by E., if the angle of obliquity be but a half point, fifty-six minutes if it be two points, thirty-two minutes if it be four, and twenty-three minutes if it be six points. I think four points should be preferred when practicable, especially *in the presence of an enemy*, since the vessels while forming would be in direct echelon, and, by "slowing," the wing nearest the forming column could be thrown into *double echelon*, thus making, with the vessels already in column, the three sides of a square *impossible to penetrate*.

---

* As the arcs described by the different vessels in coming to the oblique course will not be exactly the same, this is not mathematically correct, but it will be found *practically* so, provided the vessels continue at full speed until on the oblique course.

2. The fleet being in line heading N., to form it into columns of vessels from the right of divisions, preserving the original direction.

1st. Method.

FIG. 19.

The commander-in-chief signals :

*From the right of divisions, form columns of vessels—fleet N. E.—right vessels N.**

Flag-ships of divisions signal : *Division N. E.—right vessel N.**

The vessels on the right of divisions keep their course ; the other vessels steer N. E., and come into column in the wake of their leaders.

2d. Method.

FIG. 20.

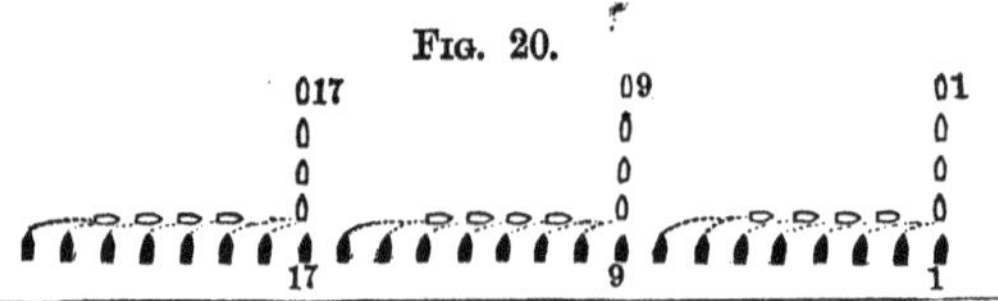

* Right vessels' (1, 9, and 17) distinguishing pennant must here be hoisted over compass signal N.

The commander-in-chief signals :

*From the right of divisions, form columns of vessels—fleet E.—right vessels N.**

Flag-ships of divisions signal : *Division E.—right vessel N.**

The vessels on the right of divisions keep their course ; the other vessels steer E , and come into column in the wake of their leaders.

It is evident that the fleet can be formed into columns of vessels from the left of divisions, and from the right and left of squadrons, preserving the original direction, according to the same principles.

3. The fleet being in line, heading N., to form it into double column, from the right, preserving the original direction.

1st. Method.

FIG. 21.

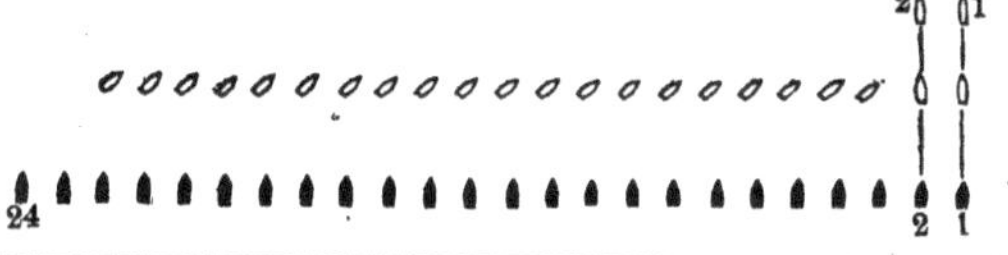

* Distinguishing pennants of 1, 9, and 17, to be hoisted over compass signal N.

The commander-in-chief signals:

*From the right of fleet, form double column—fleet N. E.—right vessels N.**

Flag-ship of van division: *Division N. E.—right vessels* N.*

Flag-ships of centre and rear divisions signal: *Division N. E.*

The two vessels on the right of the fleet keep their course; the others steer N. E; the odd-numbered vessels coming into column in the wake of one, the even-numbered in the wake of two.

2*d. Method.*

FIG. 22.

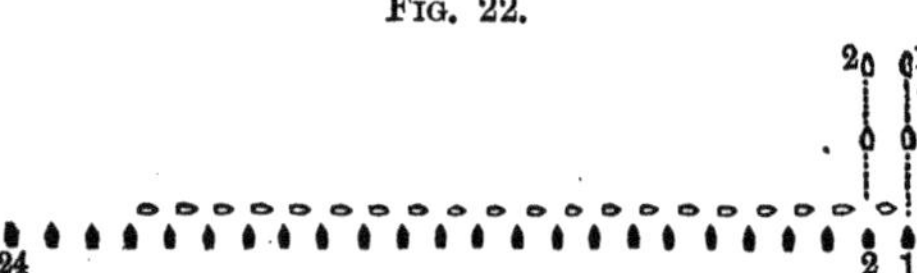

The commander-in-chief signals:

*From the right of fleet form double column—fleet E.—right vessels N.**

* Distinguishing pennants of 1 and 2 hoisted over compass signal N.

Flag-ship of van division : *Division E.—right vessels N.**

Flag-ships of centre and rear divisions signal : *Division E.*

The two vessels on the right of the fleet keep their course ; the others steer E., and come into column as before.

It is evident that the fleet can be formed into double column from the left, and into triple column, column of fours,† etc., etc., from the right and left, according to the same principles.

4. The fleet being in line, heading N., to form it into double columns, from the right of divisions, preserving the original direction.

1st Method.

FIG. 23.

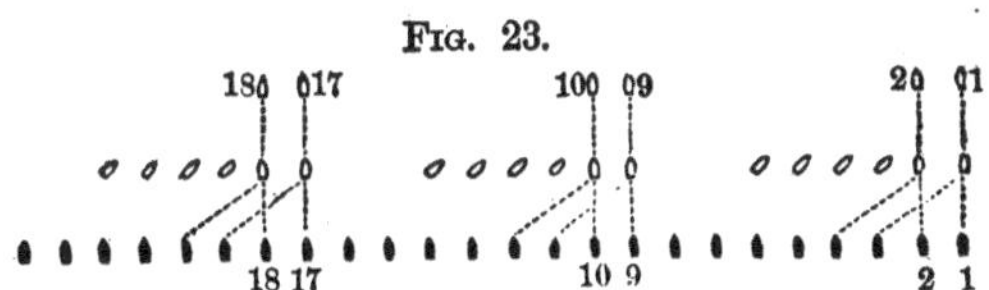

The commander-in-chief signals :

From the right of divisions—form double columns—fleet N. E.—right vessels N.

* Distinguishing pennants of one and two must be hoisted over compass signal N.

† In this case column of squadrons, as the divisions consist of eight vessels each, forming two squadrons.

Flag-ships of divisions signal : *Division N. E. —right vessels N.*

The two vessels on the right of divisions keep their course ; the others steer N. E. and come into column in the wake of their leaders.

2d Method.

FIG. 24.

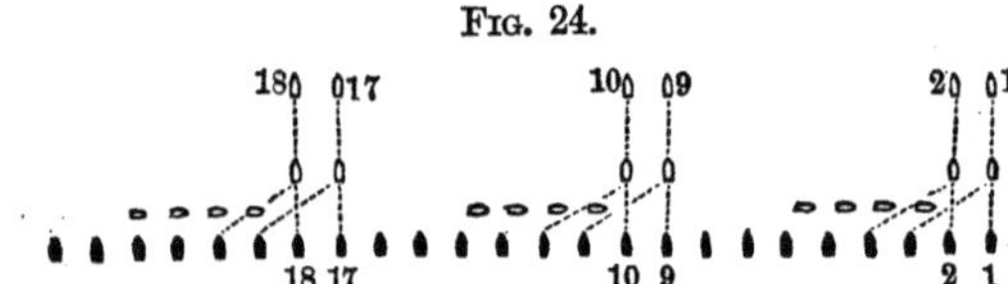

The commander-in-chief signals :

From the right of divisions—form double column —fleet E.—right vessels N.

Flag-ships of divisions signal : *Division E.— right vessels N.*

The two vessels on the right of divisions keep their course ; the others steer E. and come into column, as before.

It is evident that the fleet can be formed into double columns from the left of divisions and into triple column, column of fours, etc., from the right and left of divisions, according to the same principles.

5. The fleet being in line, heading N., to form it into column of divisions from the right, preserving the original direction.

1st Method.

FIG. 25.

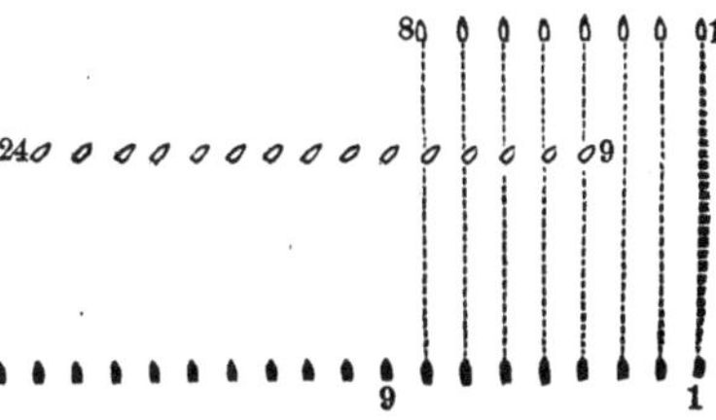

The commander-in-chief signals :

*From the right of fleet, form column of divisions —fleet N. E.—right division N.**

Flag-ship of van division : *Division N.*

Flag-ships of centre and rear divisions signal : *Division N. E.*

The van division keeps its course ; the other divisions steer N. E.; the centre division coming into column in the wake of the van, and the rear in the wake of the centre.

---

* Right division's distinguishing pennant must be hoisted above compass signal N.

2d Method.

FIG. 26.

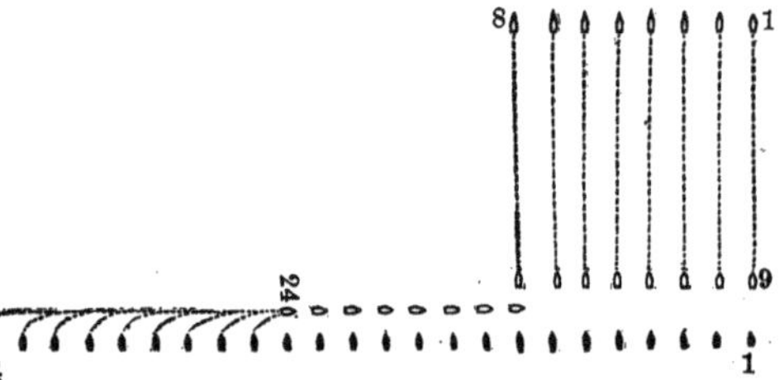

The commander-in-chief signals :

*From the right of fleet, form column of divisions —fleet E.—right division N.**

Flag-ship of van division : *Division N.*

Flag-ships of centre and rear divisions signal : *Division E.*

The van division keeps its course ; the other divisions steer E., and come into column, as before.

It is evident that the fleet can be formed into column of squadrons from the right and into column of divisions or squadrons from the left, according to the same principles.

---

* Right division's distinguishing pennant must be hoisted above compass signal N.

6. The fleet being in line, heading N., to form it into columns of squadrons from the right of divisions, preserving the original direction.

1st *Method.*

FIG. 27.

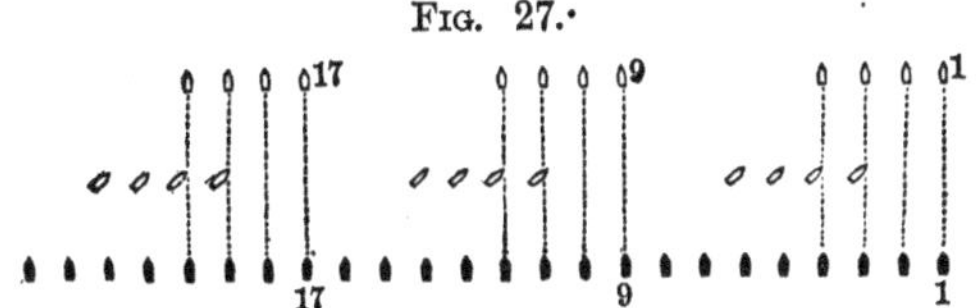

The commander-in-chief signals:

From the right of divisions, form column of squadrons—fleet N.E.—right squadrons N.*

The flag-ships of divisions signal: *Division N.E.—right squadron N.**

The squadrons on the right of divisions keep their course; the others steer N.E., and come into column in the wake of their leaders.

2d *Method.*

FIG. 28.

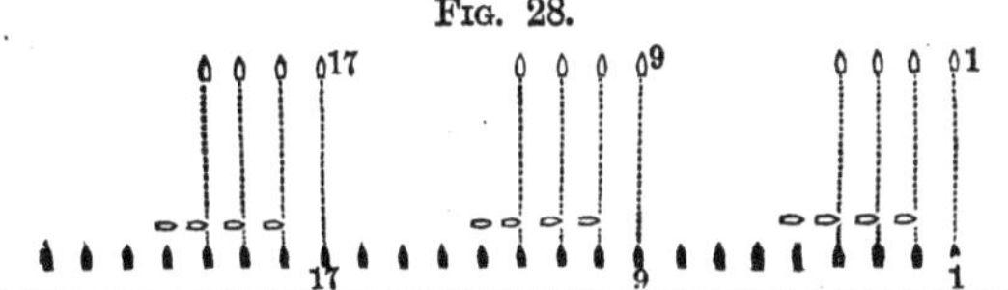

* Right squadrons' distinguishing pennants must be hoisted above compass signal *N.*

The commander-in-chiefs signals:

From the right of divisions, form column of squadrons—fleet E.—right squadrons N.

The flag-ships of divisions signal: *Division E. —right squadron N.*

The squadrons on the right of divisions keep their course; the others steer E., and come into column as before.

It is evident that the fleet can be formed into columns of squadrons from the left of divisions, according to the same principles.

7. The fleet being in line, heading North, to form it into double column from the centre, preserving the original direction.

1*st Method.*

FIG. 29.

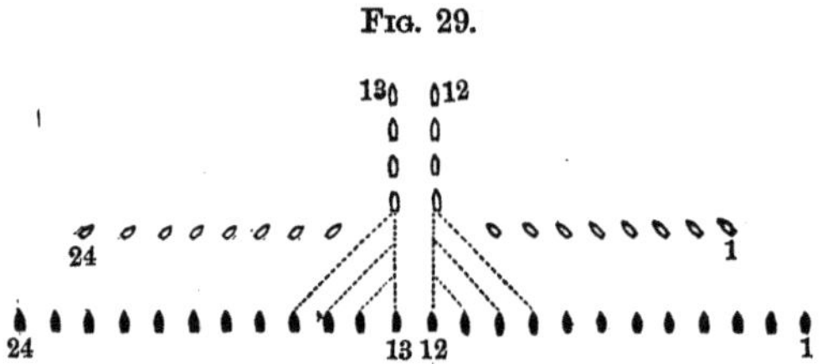

The commander-in-chief signals :

*From the centre of the fleet, form double column —right wing N.W.—left wing N.E.—centre vessels N.**

The flag-ship of van division : *Division N.W.*

The flag-ship of rear division : *Division N.E.*

The flag-ship of centre division : *Right vessels*† *N.W.; left vessels*† *N.E.; centre vessels*† *N.*

The two centre vessels keep their course ; the vessels of the right wing steer N.W.; those of the left wing N.E. : the former coming into column in the wake of the right centre, the latter in the wake of the left centre vessel.

2d *Method.*

FIG. 30.

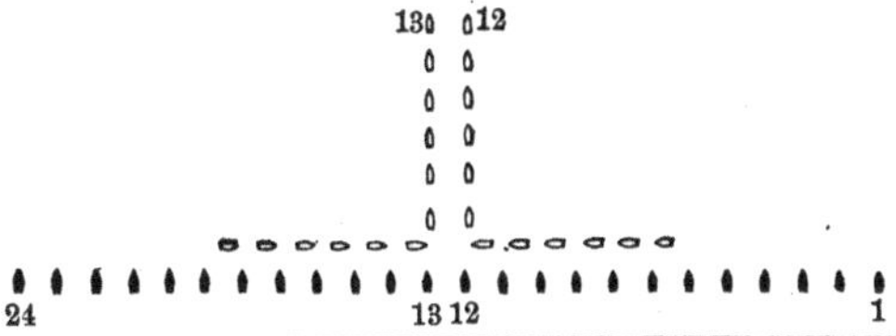

13 12

24 13 12 1

* Centre vessels distinguishing pennants must be hoisted over compass signal *N.*

† Distinguishing pennants must be hoisted above compass signal.

The commander-in-chief signals:

From the centre of the fleet, form double column —right wing W.; left wing E.; centre vessels N.*

The flag-ship of van division: *Division W.*

The flag-ship of rear division: *Division E.*

The flag-ship of centre division: *Right vessels† W.; left vessels E.†; centre vessels† N.*

The two centre vessels keep their course; the vessels of the right wing steer W.; those of the left wing steer E., and all come into column as before.

8. The fleet being in line, heading N., to form it into column of divisions from the centre, preserving the original direction.

1st Method.

FIG. 31.

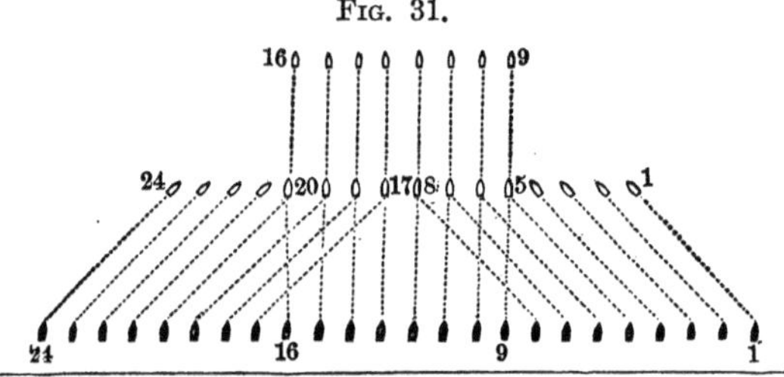

* Centre vessels' distinguishing pennants must be hoisted over compass signal *N.*

† Distinguishing pennants must be hoisted above compass signal.

The commander-in-chief signals:

From the centre of fleet, form column of divisions—right division N. W.—left division* N. E.—centre division* N.*

The flag-ship of van division : *Division N. W.*

The flag-ship of rear division : *Division N. E.*

The flag-ship of centre division: *Division N.*

The centre division keeps its course ; the van division steers N. W., the rear division N. E.; the next to van and next to rear squadrons come into column in the wake of the right and left centre squadrons, respectively, and the van and rear squadrons in like manner, in the wake of the next to van and next to rear squadrons.

2d Method.

Fig. 32.

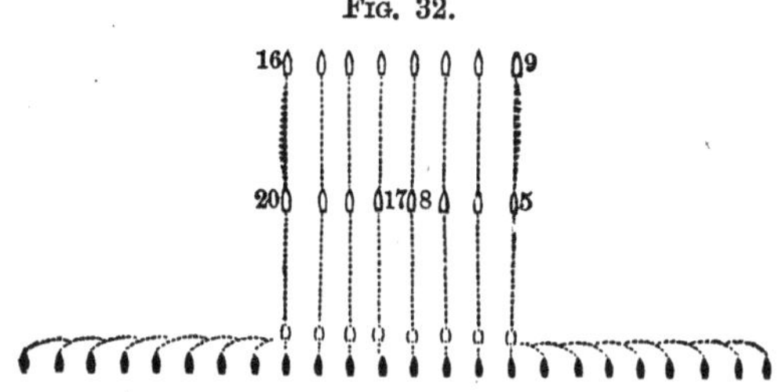

* Distinguishing pennants of divisions must be hoisted above compass signals.

The commander-in-chief signals :

From the centre of fleet, form column of divisions —right division W.—left division E.—centre division N.

The flag-ship of van division : *Division W.*

The flag-ship of rear division : *Division E.*

The flag-ship of centre division : *Division N.*

The centre division keeps its course ; the van division steers W., and the rear division E., and both come into column, as before.

9. The fleet being in line, heading N., to form it into double columns from the centre of divisions, preserving the original direction.

1*st Method.*

FIG. 33.

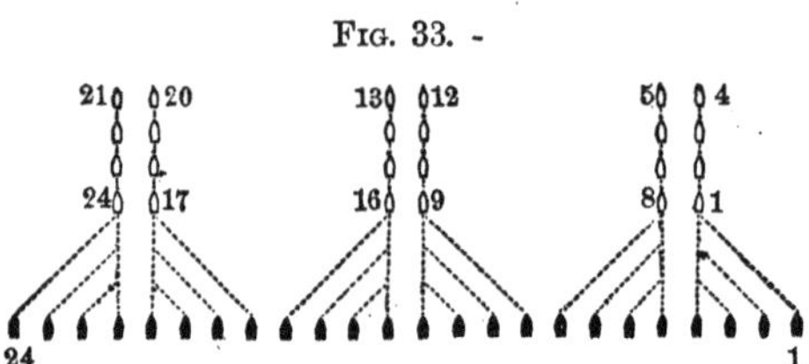

The commander-in-chief signals :

From the centre of divisions, form double columns—right vessels N. W.—left vessels N. E.—centre vessels N.

Flag-ships of divisions : *Right vessels N. W.**—*left vessels N. E.**—*centre vessels N.**

The two centre vessels of divisions keep their course; the vessels on the right of them in each division steer N. W.; those on the left N. E.; the former coming into column in the wake of the right centre, the latter in the wake of the left centre vessels.

2*d. Method.*

FIG. 34.

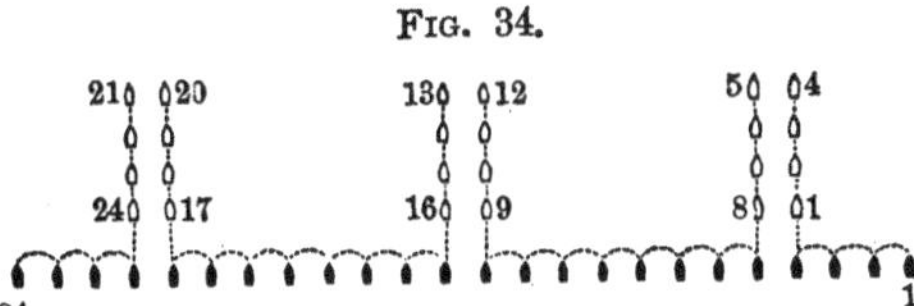

The commander-in-chief signals :

From the centre of divisions, form double column —right vessels W.—left vessels E.—centre vessels N.

The flag-ships of divisions signal : *Right vessels W.—left vessels E.—centre vessels N.*

The two centre vessels of divisions keep their course ; the vessels on the right of them in each division steer W. ; those on the left E. ; and all come into column as before. It is evident that

---

* Distinguishing pennants must be hoisted above compass signals.

the fleet can be formed into double columns from the centre of squadrons, according to the same principles.

10. The fleet being in line, heading N., to form it into column of vessels from the right, at right angles to the original direction.

FIG. 35.

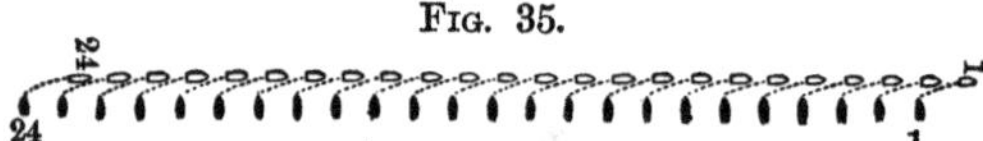

The commander-in-chief makes general signal *E.*, which is repeated by the division commanders.

All come E. in the wake of the leader of the column, and continue onward.

It is evident that the fleet can be formed into column from the left, at right angles to the original direction, according to the same principles.

11. The fleet being in line, heading N., to form it into double column from the right, at right angles to the original direction.

1*st. Method.*

FIG. 36.

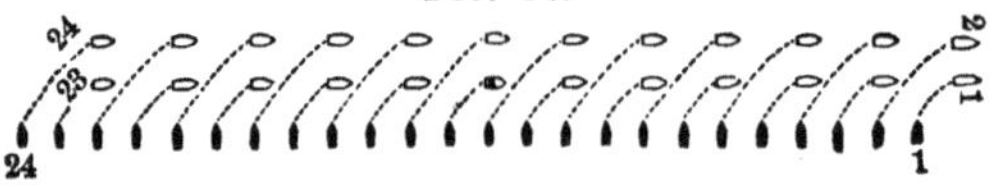

The commander-in-chief signals :

Fleet—by twos—wheel to E.

Flag-ships of divisions signal: *Division—by twos—wheel to E.*

The fleet steers N. E. ; the pivot-vessels (the odd-numbered vessels in this case) under steerage way, the others maintaining their speed. When the leader of the column—*the right pivot-vessel*—finds her consort bearing from her on a line perpendicular to the course given by the commander-in-chief (N. is the bearing in this case), she breaks the stops of compass signal E., which has been "rounded up at the fore, when all come together E., and the column is formed.*

2d. Method.

The commander-in-chief makes general signal *E.*, which is repeated by the division commanders.

Executed as in manœuvre 10, Fig. 35, and afterward:

Fleet—form double column—left oblique.

Executed as in manœuvre 27, Fig. 56.

It is evident that the fleet can be formed into triple column, column of fours, etc., from the

* Column will be formed in one minute and forty-five seconds, when the fleet is moving ten knots an hour.

right, at right angles to the original direction, as also into double column, triple column, etc., etc., from the left, according to the same principles.

12. The fleet being in line, heading N., to form it into column of divisions from the right, at right angles to the original direction.

1*st Method.*

Fig. 37.

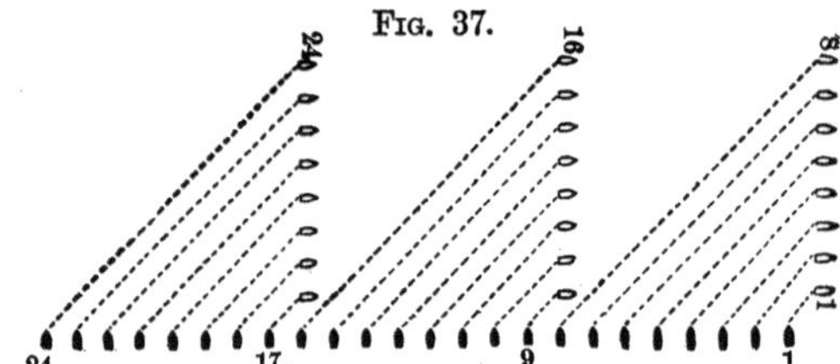

The commander-in-chief signals:

By divisions—wheel to E.

The flag-ships of divisions signal : *Division—wheel to E.*

The fleet steers N.E., the pivot-vessels (1, 9, and 17) under steerage-way; those farthest from the pivots at full speed, and the intervening vessels* as shown by annexed table. When

* Or they may proceed at full speed and slow to steerage-way, so soon as they bring the pivot-vessel to bear S.

the pivot-vessels find their consorts* bearing from them on a line perpendicular to the given course, they hoist the *position pennant,* when all come together E., and the column is formed.

NOTE TO 1ST METHOD.—The divisions will change front eight points, by wheeling, in 12½ minutes, supposing the speed of the pivot-vessels to be four knots, that of the vessels farthest from the pivots ten knots, and that of the intervening vessels as shown by the annexed table. I think the change of front by the 2d method infinitely preferable :

Pivot.	Next to Pivot.	3d.	4th.	5th.	6th.	7th.	8th.
4.0	4.9	5.7	6.6	7.4	8.3	9.2	10.0

2d Method.

FIG. 38.†

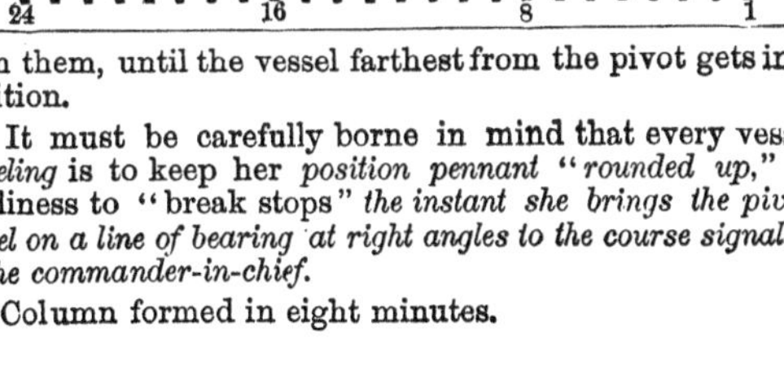

from them, until the vessel farthest from the pivot gets into position.

* It must be carefully borne in mind that every vessel *wheeling* is to keep her *position pennant "rounded up,"* in readiness to "break stops" *the instant she brings the pivot-vessel on a line of bearing at right angles to the course signalled by the commander-in-chief.*

† Column formed in eight minutes.

The fleet is first formed into columns of vessels from the left of divisions, as prescribed in manœuvre 2, after which the commander-in-chief makes general signal *E.*, which is repeated by the divisional commanders, and the column is formed. Should the fleet be formed into column of vessels from the right of divisions, in this case, and then into columns of divisions to the right, it would be steaming in natural order, that is with the van division leading, while its divisions would be in order reversed, with their *right squadrons on the left.*

3*d* *Method.*

The fleet is formed into column of vessels, as in manœuvre 10, Fig. 35, and afterward into column of divisions as in manœuvre 29, Fig. 58.

It is evident that the fleet can be formed into column of divisions from the left, at right angles to the original direction, according to the same principles.

13. The fleet being in line, heading N., to form it into column of vessels from the right, on any course from N. to E.

Suppose, for example, it be required to form the column to the N.N.E.

1st Method.

FIG. 39.

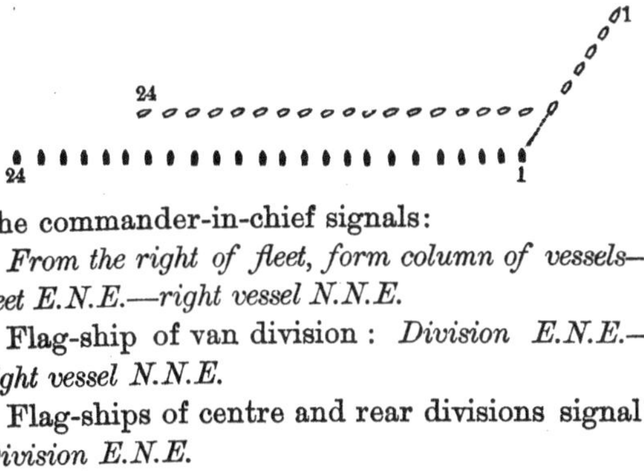

24 24 1 1

The commander-in-chief signals:

From the right of fleet, form column of vessels—fleet E.N.E.—right vessel N.N.E.

Flag-ship of van division: *Division E.N.E.—right vessel N.N.E.*

Flag-ships of centre and rear divisions signal: *Division E.N.E.*

The vessel on the right of the fleet keeps N.N.E.; the other vessels steer E.N.E.; two coming N.N.E. when in the wake of one, three in the wake of two, and so on to the last vessel.

2d Method.

FIG. 40.

The commander-in-chief signals :

From the right of fleet, form column of vessels—fleet E.—right vessel N.N.E.

Flag-ship of van division: *Division E.—right vessel N.N.E.*

Flag-ships of centre and rear divisions signal: *Division E.*

The vessel on the right of the fleet steers N.N.E. ; the other vessels keep E.; two coming N.N.E. in the wake of one, three in the wake of two, and so on to the last vessel.

It is evident that the fleet can be formed into column from the left, on any course from N. to W., according to the same principles.

NOTE.—The following tables show the rate of speed of the vessels, calculated for each half-point of obliquity from the *course signalled by the commander-in-chief for the leading or guide vessel or vessels:*

TABLE C.

Angle of Obliquity in Degrees and parts of a Degree.	Angle of Obliquity in Points and Half Points.	Speed of Obliquing Vessels in Knots and Tenths.	Speed of Leading or Guide Vessel (or Vessels) in Knots., Steering N. by E.
° ′ ″			
(a) 5 37 30	½ point N. by E. ½ E.....	9.3	10
11 15 00	1 point N. N. E.........	8.8	10
16 52 30	1½ points N. N. E. ½ E...	8.4	10
(b) 22 30 00	2 points N. E. by N......	8.1	10
28 07 30	2½ points N. E. ½ N......	7.9	10
33 45 00	3 points N. E...........	7.8	10
39 22 30	3½ points N. E. ½ E.......	7.7	10
(c) 45 00 00	4 points N. E. by E......	7.8	10
50 37 30	4½ points N. E. by E. ½ E.	7.9	10
56 15 00	5 points E. N. E.........	8.1	10
61 52 30	5½ points E. N. E. ½ E....	8.4	10
(d) 67 30 00	6 points E. by N.........	8.8	10
73 07 30	6½ points E. ½ N.........	9.3	10

TABLE D.

Angle of Obliquity in Degrees and parts of a Degree.	Angle of Obliquity in Points and Half Points.	Speed of Obliquing Vessels in Knots and Tenths.	Speed of Leading or Guide Vessel (or Vessels) in Knots, Steering N. N. E.
(e) ° ′ ″ 5 37 30	½ point N. N. E. ½ E.. ..	9.4	10
11 15 00	1 point N. E. by N.... ..	9.0	10
16 52 30	1½ point N. E. ½ N.......	8.7	10
22 30 00	2 points N. E......	8.5	10
28 07 30	2½ points N. E. ½ E......	8.4	10
33 45 00	3 points N. E. by E......	8.3	10
39 22 30	3½ points N. E. by E. ½ E.	8.4	10
45 00 00	4 points E. N. E.........	8.5	10
50 37 30	4½ points E. N. E. ½ E. ..	8.7	10
56 15 00	5 points E. by N....	9.0	10
61 52 30	5½ points E. ½ N.........	9.4	10

TABLE E.

Angle of Obliquity in Degrees and parts of a Degree.	Angle of Obliquity in Points and Half Points.	Speed of Obliquing Vessels in Knots and Tenths.	Speed of Leading or Guide Vessel (or Vessels) in Knots, Steering N. E. by N.
(*f*) 5° 37′ 30″	½ point N. E. ½ N........	9.5	10
11 15 00	1 point N. E...........	9.2	10
16 52 30	1½ points N. E. ½ E......	9 0	10
22 30 00	2 points N. E. by E......	8.9	10
28 07 30	2½ points N. E. by E. ½ E.	8.8	10
33 45 00	3 points E. N. E.........	8.9	10
39 22 30	3½ points E. N. E. ½ E...	9.0	10
45 00 00	4 points E. by N........	9.2	10
50 37 30	4½ points E. ½ N........	9.5	10

TABLE F.

Angle of Obliquity in Degrees and parts of a Degree.	Angle of Obliquity in Points and Half Points.	Speed of Obliquing Vessels in Knots and Tenths.	Speed of Leading or Guide Vessel (or Vesels) in Knots, Steering N. E.
(*g*) 5° 37′ 30″	½ point N. E. ½ E.........	9.6	10
11 15 00	1 point N. E. by E.......	9.4	10
16 52 30	1½ points N. E. by E. ½ E.	9.3	10
22 30 00	2 points E. N. E.........	9.2	10
28 07 30	2½ points E. N. E. ½ E...	9.3	10
33 45 00	3 points E. by N........	9.4	10
39 22 30	3½ points E. ½ N........	9 6	10

TABLE G.

Angle of Obliquity in Degrees and parts of a Degree.	Angle of Obliquity in Points and Half Points.	Speed of Obliquing Vessels in Knots and Tenths.	Speed of Leading or Guide Vessel (or Vessels) in Knots, Steering N. E. by E.
(*h*) 5° 37′ 30″	½ point N. E. by E. ½ E..	9.8	10
11 15 00	1 point E. N. E....	9.6	10
16 52 30	1½ points E. N. E. ½ E...	9.6	10
22 30 00	2 points E. by N........	9.6	10
28 07 30	2½ points E. ½ N..........	9.8	10

TABLE H.

Angle of Obliquity in Degrees and parts of a Degree.	Angle of Obliquity in Points and Half Points.	Speed of Obliquing Vessels in Knots and Tenths.	Speed of Leading or Guide Vessel (or Vessels) in Knots, Steering E. N. E.
(*i*) 5° 37′ 30″	½ point E. N. E. ½ E.....	9.9	10
11 15 00	1 point E. by N...... ...	9.8	10
16 52 30	1½ points E. ½ N....	9.9	10

	H.	Min.
(a) column will be formed in.............	2	56
(b) " " "		55
(c) " " "		30
(d) " " "		20
(e) " " "	2	43
(f) " " "	2	26
(g) " " "	2	02
(h) " " "	1	34
(i) " " "	1	04

See remarks on page 18.

14. The fleet being in line, heading N., to form it into column of vessels from the right, on any course from N. to W.

To form the column to the N.N.W.

1st Method.

FIG. 41.

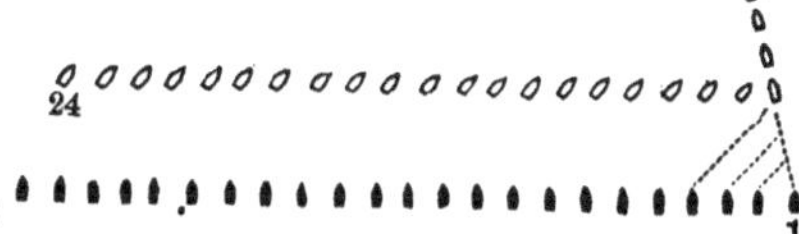

The commander-in-chief signals:

From the right of fleet, form column of vessels—fleet N.E.—right vessel N.N.W.

Flag-ship of van division: *Division N.E.—right vessel N.N.W.*

Flag-ships of centre and rear divisions signal: *Division N.E.*

The vessel on the right of the fleet steers N.N.W.; the other vessels N.E., two coming N.N.W., when in the wake of one, three in the wake of two, and so on to the last vessel.

2d Method.

FIG. 42.

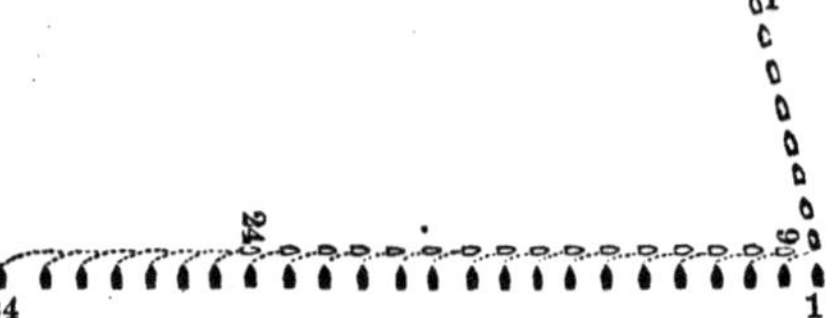

The commander-in-chief signals:

From the right of fleet, form column of vessels—fleet E.—right vessel N.N.W.

Flag-ship of van division: *Division E.—right vessel N.N.W.*

Flag-ships of centre and rear divisions signal: *Division E.*

The vessel on the right of the fleet steers N.N.W.; the other vessels E., and come into column as before.

It is evident that the fleet can be formed into column of vessels from the left, on any course from N. to E., according to the same principles.

The following tables show the rate of speed of

the vessels calculated for each half-point of obliquity from the course signalled by the commander-in-chief, for the leading or guide vessel (*or vessels*).

TABLE I.

Angle of Obliquity in Degrees and Parts of a Degree.	Angle of Obliquity in Points and Half Points.	Speed of Obliquing Vessels in Knots and Tenths.	Speed of Leading or Guide Vessel (or Vessels) in Knots, Steering N. by W.
° ′ ″ 5 37 30	½ point N. ½ W..........	9.0	10
11 15 00	1 point North...........	8.2	10
16 52 30	1½ points N. ½ E....	7.8	10
22 30 00	2 points N. by E....... ..	7.2	10
28 07 30	2½ points N. by E. ½ E...	7.0	10
33 45 00	3 points N. N. E.........	6.6	10
39 22 30	3½ points N. N. E. ½ E...	6.5	10
45 00 00	4 points N. E. by N......	6.4	10
50 37 30	4½ points N. E. ½ N.	6.3	10
56 15 00	5 points N. E......... ...	6.4	10
61 52 30	5½ points N. E. ½ E......	6.5	10
67 30 00	6 points N. E. by E......	6.6	10
73 07 30	6½ points N. E. by E. ½ E.	7.0	10
78 45 00	7 points E. N. E.	7.2	10
84 22 30	7½ points N. E. ½ E......	7.8	10
90 00 00	8 points E. by N..	8.2	10
95 37 30	8½ points E. ½ N.........	9.0	10

TABLE J.

Angle of Obliquity in Degrees and Parts of a Degree.	Angle of Obliquity in Points and Half Points.	Speed of Obliquing Vessels in Knots and Tenths.	Speed of Leading or Guide Vessel (or Vessels) in Knots, Steering N. N. W.
° ′ ″			
5 37 30	½ point N. by W. ½ W...	8.8	10
11 15 00	1 point N. by W........	7.9	10
16 52 30	1½ points N. ½ W........	7.4	10
22 30 00	2 points North	6.7	10
28 07 30	2½ points N. ½ E........	6.4	10
33 45 00	3 points N. by E....	6.0	10
39 22 30	3½ points N. by E. ½ E...	5.9	10
45 00 00	4 points N. N. E........	5.7	10
50 37 30	4½ points N. N. E. ½ E...	5.6	10
56 15 00	5 points N. E. by N.. ...	5.6	10
61 52 30	5½ points N. E. ½ N......	5.6	10
67 30 00	6 points N. E........	5.7	10
73 07 30	6½ points N. E. ½ E......	5.9	10
78 45 00	7 points N. E. by E......	6.0	10
84 22 30	7½ points N. E. by E. ½ E.	6.4	10
90 00 00	8 points E. N. E....	6.7	10
95 37 30	8½ points E. N. E. ½ E...	7.4	10
101 15 00	9 points E. by N...... ..	7.9	10
106 52 30	9½ points E. ½ N...... ..	8.8	10

TABLE K.

Angle of Obliquity in Degrees and Parts of a Degree.	Angle of Obliquity in Points and Half Points.	Speed of Obliquing Vessels in Knots and Tenths.	Speed of Leading or Guide Vessel (or Vessels) in Knots, Steering N.W. by N.
° ′ ″			
5 37 30	½ point N. N. W. ½ W....	8.5	10
11 15 00	1 point N. N. W.........	7.4	10
16 52 30	1½ points N. by W. ½ W..	6.7	10
22 30 00	2 points N. by W........	6.1	10
28 07 30	2½ points N. ½ W....	5.7	10
33 45 00	3 points North....	5.3	10
39 22 30	3½ points N. ½ E.........	5.1	10
45 00 00	4 points N. by E....	4.9	10
50 37 30	4½ points N. by E. ½ E...	4.8	10
56 15 00	5 points N. N. E.........	4.7	10
61 52 30	5½ points N. N. E. ½ E. ..	4.6	10
67 30 00	6 points N. E. by N.....	4.7	10
73 07 30	6½ points N. E. ½ N......	4.8	10
78 45 00	7 points N. E......	4.9	10
84 22 30	7½ points N. E. ½ E......	5.1	10
90 00 00	8 points N. E. by E.	5 3	10
95 37 30	8½ points N. E. by E. ½ E.	5.7	10
101 15 00	9 points E. N. E..	6.1	10
106 52 30	9½ points E. N. E. ½ E...	6.7	10
112 30 00	10 points E. by N.......	7.4	10
118 07 30	10½ points E. ½ N.	8.5	10

TABLE L.

Angle of Obliquity in Degrees and Parts of a Degree.	Angle of Obliquity in Points and Half Points.	Speed of Obliquing Vessels in Knots and Tenths.	Speed of Leading or Guide Vessel (or Vessels) in Knots, Steering N. W.
° ′ ″ 5 37 30	½ point N. W. ½ N.... ...	8.1	10
11 15 00	1 point N. W. by N......	6.9	10
16 52 30	1½ points N. N. W. ½ W..	6.1	10
22 30 00	2 points N. N. W........	5.4	10
28 07 30	2½ points N. by W. ½ W..	4.9	10
33 45 00	3 points N. by W........	4.6	10
39 22 30	3½ points N. ½ W........	4.3	10
45 00 00	4 points North	4.1	10
50 37 30	4½ points N. ½ E........	4.0	10
56 15 00	5 points N. by E.........	3.9	10
61 52 30	5½ points N by E. ½ E....	3.8	10
67 30 00	6 points N. N. E.........	3.8	10
73 07 30	6½ points N. N. E. ½ E. ..	3.8	10
78 45 00	7 points N. E. by N.. ...	3.9	10
84 22 30	7½ points N. E. ½ N.. ...	4.0	10
90 00 00	8 points N. E..........	4.1	10
95 37 30	8½ points N. E. ½ E.....	4.3	10
101 15 00	9 points N. E. by E......	4.6	10
106 52 30	9½ points N. E. by E. ½ E.	4.9	10
112 30 00	10 points E. N. E........	5.4	10
118 07 30	10½ points E. N. E. ½ E..	6.1	10
123 45 00	11 points E. by N.... ...	6.9	10
129 22 30	11½ points E. ½ N........	8.1	10

TABLE M.

Angle of Obliquity in Degrees and Parts of a Degree.	Angle of Obliquity in Points and Half Points.	Speed of Obliquing Vessels in Knots and Tenths.	Speed of Leading or Guide Vessel (or Vessels) in Knots, Steering N. W. by W.
° ′ ″			
5 37 30	½ point N. W. ½ W.......	7.6	10
11 15 00	1 point N. W............	6.2	10
16 52 30	1½ points N. W. ½ N.. ...	5.5	10
22 30 00	2 points N. W. by N.....	4.6	10
28 07 30	2½ points N. N. W. ½ W..	4.3	10
33 45 00	3 points N. N. W........	3.8	10
39 22 30	3½ points N. by W. ½ W..	3.6	10
45 00 00	4 points N. by W........	3.3	10
50 37 30	4½ points N. ½ W........	3.2	10
56 15 00	5 points North....	3.0	10
61 52 30	5½ points N. ½ E.........	3.0	10
67 30 00	6 points N. by E....	2.9	10
73 07 30	6½ points N. by E. ½ E...	2.9	10
78 45 00	7 points N. N. E....	2.9	10
84 22 30	7½ points N. N. E. ½ E...	3.0	10
90 00 00	8 points N. E. by N......	3.0	10
95 37 30	8½ points N. E. ½ N.... ..	3.2	10
101 15 00	9 points N. E...........	3.3	10
106 52 30	9½ points N. E. ½ E......	3.6	10
112 30 00	10 points N. E. by E.....	3.8	10
118 07 30	10½ points N. E. by E. ½ E.	4.3	10
123 45 00	11 points E. N. E........	4.6	10
129 22 30	11½ points E. N. E. ½ E..	5.5	10
135 00 00	12 points E. by N.... ...	6.2	10
140 37 30	12½ points E. ½ N.... ...	7.6	10

TABLE N.

Angle of Obliquity in Degrees and Parts of a Degree.	Angle of Obliquity in Points and Half Points.	Speed of Obliquing Vessels in Knots and Tenths.	Speed of Leading or Guide Vessel (or Vessels) in Knots, Steering W. N. W.
° ′ ″			
5 37 30	½ point N. W. by W. ½ W.	6.7	10
11 15 00	1 point N. W. by W......	5.1	10
16 52 30	1½ points N. W. ½ W.....	4.4	10
22 30 00	2 points N. W..........	3.5	10
28 07 30	2½ points N. W. ½ N......	3.2	10
33 45 00	3 points N. W. by N.....	2.8	10
39 22 30	3½ points N. N. W. ½ W..	2.6	10
45 00 00	4 points N. N. W........	2.3	10
50 37 30	4½ points N. by W. ½ W..	2.2	10
56 15 00	5 points N. by W........	2.1	10
61 52 30	5½ points N. ½ W........	2.1	10
67 30 00	6 points North..........	2.0	10
73 07 30	6½ points N. ½ E........	2.0	10
78 45 00	7 points N. by E........	1.9	10
84 22 30	7½ points N. by E. ½ E...	2.0	10
90 00 00	8 points N. N. E........	2.0	10
95 37 30	8½ points N. N. E. ½ E...	2.1	10
101 15 00	9 points N. E. by N.....	2.1	10
106 52 30	9½ points N. E. ½ N......	2.2	10
112 30 00	10 points N. E..........	2.3	10
118 07 30	10½ points N. E. ½ E.....	2.6	10
123 45 00	11 points N. E. by E....	2.8	10
129 22 30	11½ points N. E. by E. ½ E.	3.2	10
135 00 00	12 points E. N. E.......	3.5	10
140 37 30	12½ points E. N. E. ½ E..	4.4	10
146 15 00	13 points E. by N.......	5.1	10
151 52 30	13½ points E. ½ N.......	6.7	10

TABLE O.

Angle of Obliquity in Degrees and parts of a Degree.	Angle of Obliquity in Points and Half Points.	Speed of Obliquing Vessels in Knots and Tenths.	Speed of Leading or Guide Vessel (or Vessels) in Knots, Steering W. by N.
° ′ ″			
5 37 30	½ point W. N. W. ½ W....	5.0	10
11 15 00	1 point W. N. W........	3.4	10
16 52 30	1½ points N. W. by W. ½ W.	2.8	10
22 30 00	2 points N. W. by W.....	2.1	10
28 07 30	2½ points N. W. ½ W....	1.9	10
33 45 00	3 points N. W..........	1.5	10
39 22 30	3½ points N. W. ½ N.....	1.4	10
45 00 00	4 points N. W. by N.....	1.3	10
50 37 30	4½ points N. N. W. ½ W..	1.2	10
56 15 00	5 points N. N. W........	1.1	10
61 52 30	5½ points N. by W ½ W...	1.0	10
67 30 00	6 points N. by W........	1.0	10
73 07 30	6½ points N. ½ W........	1.0	10
78 45 00	7 points North..........	$.\frac{98}{100}+$	10
84 22 30	7½ points N. ½ E........	$.\frac{96}{100}+$	10
90 00 00	8 points N. by E........	$.\frac{98}{100}+$	10
95 37 30	8½ points N. by E. ½ E..	1.0	10
101 15 00	9 points N. N. E........	1.0	10
106 52 30	9½ points N. N. E. ½ E...	1.0	10
112 30 00	10 points N. E. by N.....	1.1	10
118 07 30	10½ points N. E. ½ N.....	1.2	10
123 45 00	11 points N. E..........	1.3	10
129 22 30	11½ points N. E. ½ E.....	1.4	10
135 00 00	12 points N. E. by E.....	1.5	10
140 37 30	12½ points N. E. by E. ½ E.	1.9	10
146 15 00	13 points E. N. E........	2.1	10
151 52 30	13½ points E. N. E. ½ E...	2.8	10
157 30 00	14 points E. by N.	3.4	10
163 07 30	14½ points E. ½ N........	5.0	10

15. The fleet being in line, heading N., to form it into columns of vessels, from the right of divisions, on any course from N. to E.

To form the columns to the N. E.

1*st Method.*

FIG. 43.

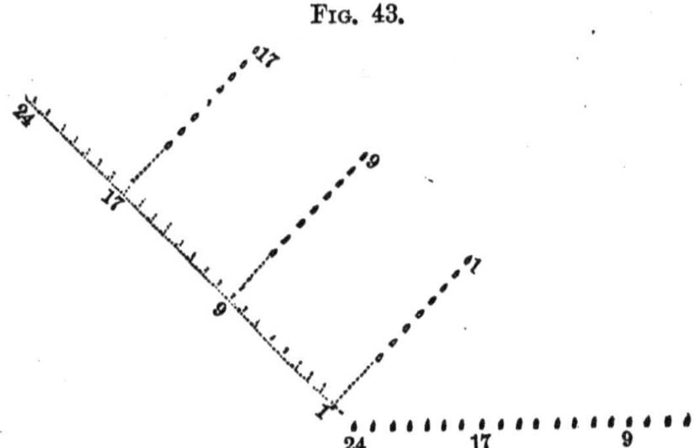

The commander-in-chief first forms the fleet into column of vessels, from the left* to the N.W. (as explained in manœuvre 13), at right angles to the course on which he intends to break

* Should the fleet be formed into column of vessels from the *right* to the N.W., and then from line into columns of vessels from the right of divisions to the N.E., it would be in reversed order, with the van division on the left, and the rear squadron of each division leading.

it into columns of vessels, from the right of divisions, then into line on that course N.E., upon general signal *N.E.*,* and completes the manœuvre by signalling, as in 2, Figs. 19 and 20.

2d Method.

FIG. 44.

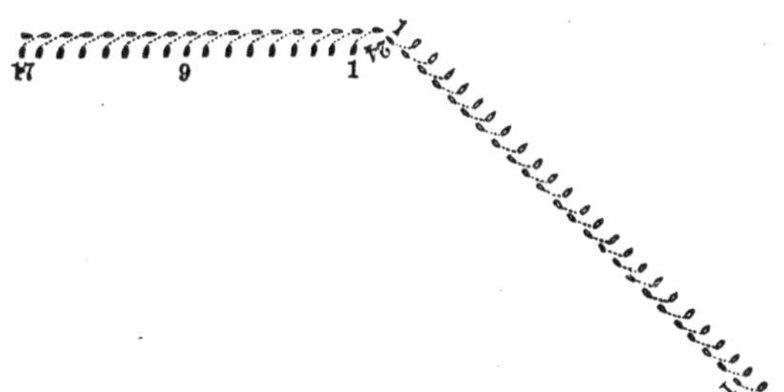

The commander-in-chief first forms the fleet into column of vessels from the right, upon general signal *E.*, then changes the direction of the column to the S.E., by signalling, as explained in 56, Fig. 87, and afterward proceeds as in the first method.

* Or he may signal *heads* of divisions N.E., as explained in manœuvre 35.

3d *Method.*

Fig. 45.

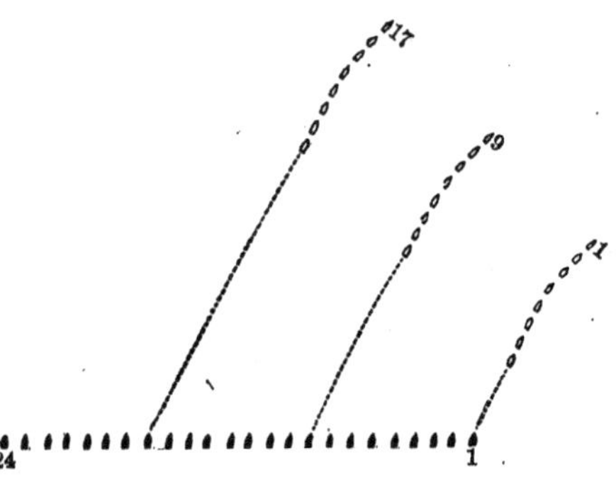

The commander-in-chief signals :

From the right of divisions, form columns of vessels—fleet E.N.E—right vessels wheel to N.E.

Flag-ships of divisions signal : *Division E.N.E.—right vessel N.N.E.*

The vessels on the right of divisions keep N.N.E.; one under steerage-way, nine* and seventeen at full speed; the other vessels steer E.N.E. at such speed as will take them into their proper places in their respective columns. (See Table D.)

---

* Or nine may preserve such speed only as will enable her to keep on a line between one and seventeen.

When nine finds one bearing from her on a line perpendicular to the course signalled by the commander-in-chief (S.E. is the bearing in this case), she hoists the *position pennant* as a guide to seventeen, and "*slows to steerage-way*," until seventeen, bringing her and one on this bearing, hoists the *position pennant*, when, upon one's "breaking the stops" of compass signal N.E., *all the leaders* of the columns keep N.E. at *full speed*; the other vessels stand on N.N.E., at full speed, until in the wake of their leaders, when they steer N.E.

4th Method.

As the 3d, except that, upon the signal of the commander-in-chief, *fleet E.*, the vessels steer E. to get into column in the wake of their leaders, steering N.N.E., and preserve, of course, the same speed as their leaders.

It is evident that the fleet can be formed into columns of vessels from the left of divisions on any course from N. to W., according to the same principles.

16. The fleet being in line, heading N., to form it into double column from the right, on any course from N. to E.

To form the column to the N.E.

1st Method.

Form the fleet into column of vessels from the right to the N.E, as in manœuvre 13, Figs. 39 and 40, and then into double column, as in 27, Fig. 56.

*2d Method.**

Fig. 46.

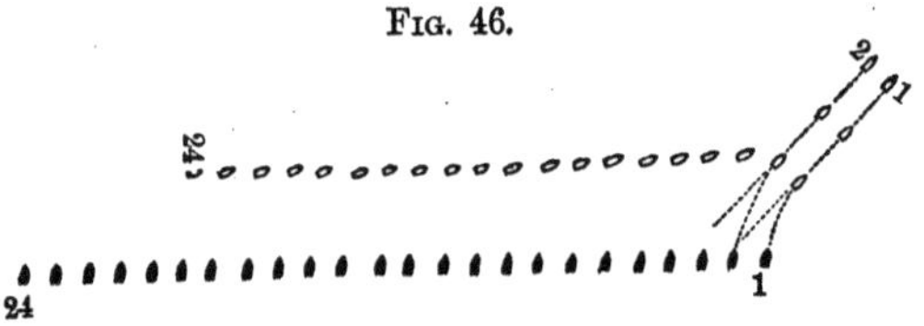

The commander-in-chief signals:

From the right of fleet, form double column—fleet E.N.E. (or E.)—right vessels wheel to N.E.

---

* This method should not be resorted to when the angle of the new course with the old is more than 45°, up to which point it will commend itself to the reader for its simplicity.

Flag-ship of van division: *Division E.N.E. (or E.)—right vessels wheel to N.E.*

Flag-ships of centre and rear divisions signal: *Division E.N.E. (or E.)*

The two vessels on the right of the fleet alter course together to N.N.E.—one, the pivot-vessel, under steerage-way; two, at full speed. When the pivot-vessel finds, *from the position pennant*, that her consort bears from her on a line perpendicular to the given course (N.W. is the bearing in this case), she breaks the stop of compass signal N.E., when both steer N.E. at full speed. The other vessels steer E.N.E.* (or E.), three coming into column in the wake of one after she is on her course N.E., four in the wake of two, etc., etc. So soon as they come into column, the *odd-numbered* vessels have to slow to steerage-way to enable their consorts to get on a line of bearing N.W. from them.

* As each pivot-vessel is delayed in turn to enable her consort (or *consorts*) to get on the proper line of bearing from her, after she gets into column the vessels will be governed, in coming into column, by the speed of the fleet when the manœuvre was signalled (as by Table F).

3d *Method.*

Fig. 47.

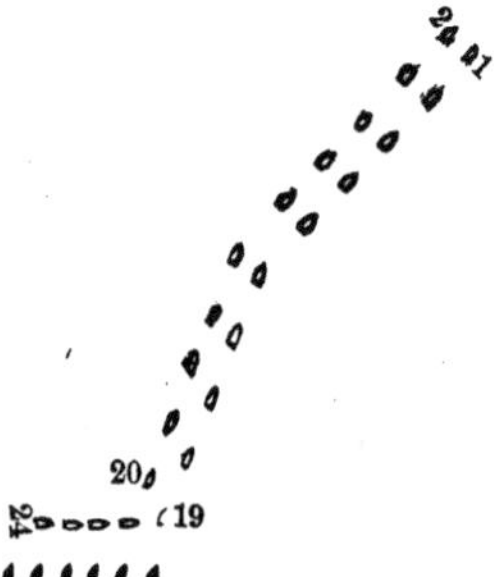

The commander-in-chief signals:

From the right of fleet, form double column—fleet E.N.E. (or E.)—right vessels wheel to N.E.

Flag-ship of van division:—*Division E.N.E. (or E.)—right vessels wheel to N.E.*

Flag-ships of centre and rear divisions signal: —*Division E.N.E. (or E.)*

The two vessels on the right of the fleet alter course together to N.N.E., at full speed; the other vessels steer E.N.E. (or E.) (at such speed as is prescribed by Table D), three coming into column in the wake of one, four in the wake of two, etc., etc.

So soon as the column is formed to the N.N.E. (or while it is still forming) one slows to steerage-way until two bears N.W. from her, when she displays the compass signal, and both she and her consort steer N.E. at full speed. The other vessels stand on N.N.E., three keeping N.E. when in the wake of one, four in the wake of two, etc.; the odd-numbered vessels* slowing successively as they come to the course, to enable their consorts to get on a line of bearing N.W. from them.

4th Method.

Form the fleet into column of vessels, from the left to the N.W. or from the right to the S.E., and then into line to the N.E. (as in manœuvre 15, Figs. 43 and 44), and afterwards into double column from the right, as in 3, Figs. 21 and 22.

It is evident that the fleet can be formed into triple column, column of fours, etc., etc., from the right, on any course from N. to E., and into double column, triple column, etc., etc., from the *left, on any course from N. to W.*, according to the same principles.

---

* Or, if in column to the N.N.E., when one slows they may then slow, and enable their consorts to attain the proper bearing.

The fleet being in line, heading N., to form it into double columns from the right of divisions, on any course from N. to E.

To form the columns to the N.E.

1st Method.

The fleet is first formed into column of vessels from the left to the N.W., or from the right to the S.E., and then into line to the N.E. (as in 15, Figs. 43 and 44), and finally into double columns from the right of divisions, as in 4, Figs. 23 and 24.

2d Method.

The fleet is first formed into columns of vessels, from the right of divisions (as in 15, Figs. 43 and 44) to the N.E., and then into double columns, as in 28, Fig. 57.

It is evident that the fleet can be formed into triple columns, columns of fours, etc., from the right of divisions, on any course from N. to E., and into double columns, triple columns, etc., etc., from the left of divisions, on any course from N. to W., according to the same principles.

18. The fleet being in line, heading N., to form it into columns of divisions, on any course from N. to W.

To form the columns to the N.W.

1st Method.

The fleet is first formed into column of vessels from the right to the N.E., or from the left to the S.W.; then into line to the N.W., upon general signal *N.W.*, and finally into columns of vessels from the right of divisions, as in 2, Figs. 19 and 20.

2d Method.

FIG. 48.

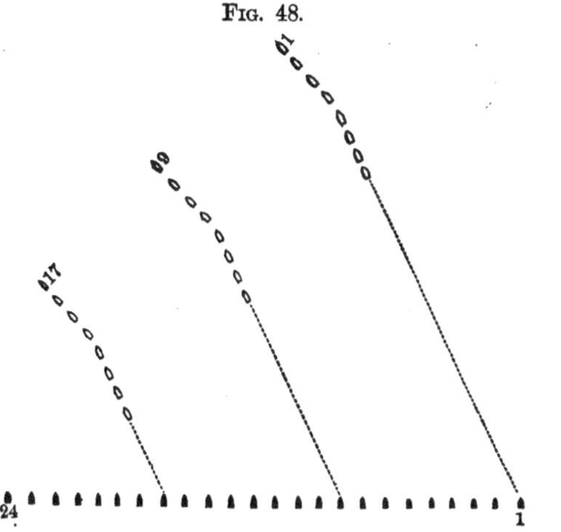

The commander-in-chief signals :

From the right of divisions, form columns of vessels—fleet N.E. (or E.)—right vessels wheel to N. W.

Flag-ships of divisions signal : *Division N.E. (or E.)—right vessels wheel to N. W.*

The vessels on the right of divisions steer N.N.W., seventeen under steerage-way,* nine and one at full speed; the other vessels keep N. E. (or E.) at such speed as will take them into their proper places in their respective columns (Table J). When nine finds seventeen bearing from her on a line perpendicular to the course signalled by the commander-in-chief (S.W. is the bearing in this case), she hoists the *position pennant,* as a guide to one, and slows to steerage-way, until one, bringing her and seventeen on this bearing, displays the position pennant, when seventeen breaks the stop of compass signal N.W., and all the leaders of the columns come to this course *at full speed.* The other vessels stand on N.N.W., *at full speed,* until they reach the wake of their leaders, when they steer N.W.

It is evident that the fleet can be formed into columns of vessels from the left of divisions, on

---

* Or nine may preserve such speed only as will enable her to keep on a line between one and seventeen.

any course from N. to E., according to the same principles.

19. The fleet being in line heading N., to form it into double column from the right, on any course from N. to W.

To form the column to the N.W.

1st Method.

Form the fleet into column of vessels from the right, as in 14, Figs. 41 and 42, and then into double column, as in 27, Fig. 56.

2d Method.

Form the fleet into column of vessels from the right to the N.E. or from the left to the S.W., then into line to the N.W., upon general signal *N.W.*, and finally into double column from the right, as in 3, Figs. 21 and 22.

3d Method.

FIG. 49.

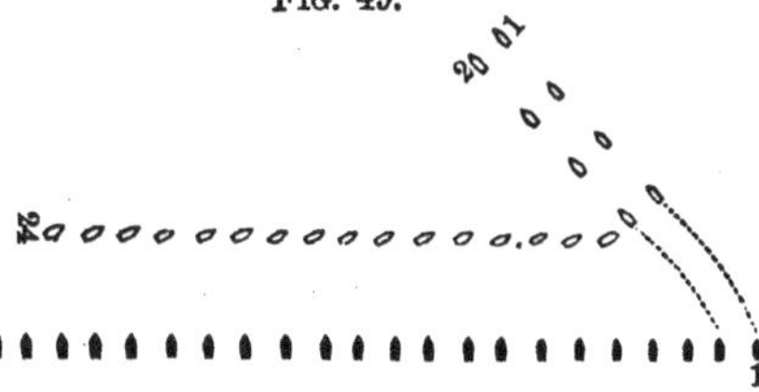

The commander-in-chief signals:

From the right of fleet, form double column—fleet N.E. (or E.)—right vessels wheel to N.W.

Flag-ship of van division: *Division N.E. (or E.)—right vessels wheel to N.W.*

Flag-ships of centre and rear divisions signal: *Division N.E. (or E.)*

The two vessels on the right of the fleet alter course together to N.N.W.—two, the pivot-vessel, under steerage-way, one at full speed. When two finds one bearing from her on a line perpendicular to the given course (N.E. is the bearing in this case), she breaks the stop of compass signal N.W., when both steer N.W. at full speed. The other vessels steer N.E. (or E.)—three coming into column in the wake of one, after the latter has commenced to steer N.W., four in the wake of two, etc., etc.

So soon as they come into column, the *even-numbered* vessels have to slow to steerage-way to enable their consorts to get on the line of bearing N. W. from them.

4th Method.

FIG. 50.

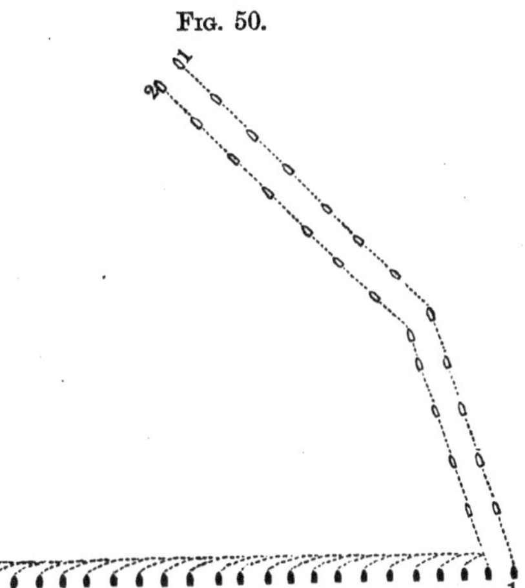

The commander-in-chief signals :

From the right of fleet, form double column—fleet N.E. (or E.)—right vessels wheel to N. W.

Flag-ship of van division : *Division N.E. (or E.)—right vessels wheel to N W.*

Flag-ships of centre and rear divisions signal : *Division N.E. (or E.)*

The two vessels on the right of the fleet alter course together to N.N.W. at full speed ; the other vessels steer N.E. (or E.)—at such speed as is prescribed by Table J; three coming into column in the wake of one, four in the wake of two, etc., etc.

So soon as the column is formed to the N.N.W. (or while it is still forming), two slows to steerage-way until one bears N.E. from her, when she displays the compass signal, and both she and her consort steer N.W. at full speed ; the other vessels stand on N.N.W., three keeping N.W. when in the wake of one, four when in the wake of two, etc., etc. ; the *even-numbered* vessels slowing successively as they come to the course to enable their consorts to get on a line of bearing N.E. from them.

It is evident that the fleet can be formed into triple column, column of fours, etc., etc., from the right, on any course from N. to W., and into double column, triple column, etc., etc., from the left, on any course from N. to E., according to the same principles.

See notes to manœuvre 16.

20. The fleet being in line, heading N., to form it into double columns from the right of divisions, on any course from N. to W.

To form the columns to the N.W.

1st Method.

The fleet is first formed into column of vessels from the right to the N.E, or from the left to the S.W., then into line to the N.W., upon general signal *N. W.*, and finally into double columns from the right of divisions, as in 4, Figs. 23 and 24.

2d Method.

The fleet is first formed into columns of vessels from the right of divisions (as in 18, Fig. 48) to the N.W., and then into double columns, as in 28, Fig. 57.

It is evident that the fleet can be formed into triple columns, columns of fours, etc., etc., from the right of divisions, on any course from N. to W., and into double column, triple column, etc., etc., from the left of divisions, on any course from N. to E., according to the same principles.

21. The fleet being in line, heading N., to change front to the right, on any course from N. to E.

To change front to N.E.

*1st Method.**

The fleet is first formed into column of vessels from the right to the S.E., or from the left to the N.W., and then into line again to the N.E., upon general signal *N.E.*

2d Method.

Fig. 51.

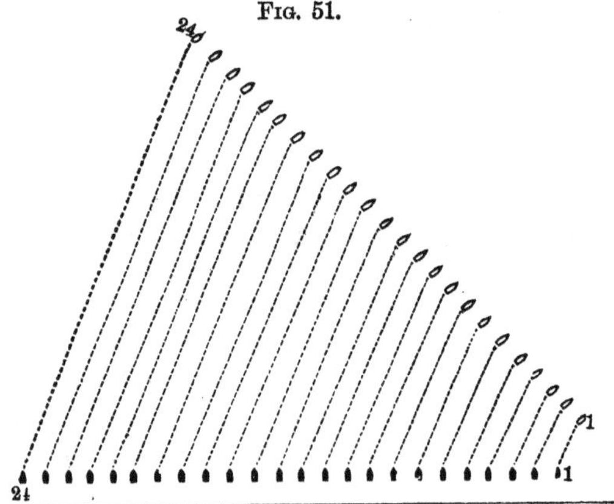

* Whenever the angle of the new course with the old exceeds six points, this is the shortest method, and I think it *always* preferable. See note to manœuvre 12.

The commander-in-chief signals :

Fleet wheel to N. E.

Flag-ships of divisions signal : *Division N.N.E.*

All the fleet alter course together to N.N.E.; one, the pivot-vessel, under steerage-way, twenty-four at full speed, and the other vessels * under such steam as will enable them to keep on a line between one and twenty-four. When the pivot-vessel finds her consorts bearing from her on a line perpendicular to the given course (N.W. is the bearing in this case), she hoists compass signal *N.E.*, when all come together to this course.

It is evident that the fleet can change front to the left, on any course from N. to W., according to the same principles, twenty-four being the pivot-vessel in this case.

22. The fleet being in line, in natural order, heading N., to change front to the rear, turning to starboard, and preserving the same order, that is, with the van squadron on the right.

The fleet is first formed into column of vessels to the right, upon general signal *E.*, as in 10, Fig.

* Or these vessels may preserve full speed, each one slowing to steerage-way, as she comes upon the bearing N.W., and hoisting the position pennant as a guide to the next vessel on her left.

35, then direction changed to W., as in 56, Fig. 87 (see note to 70), and afterward into line, heading S., upon general signal S.

It is evident that the fleet can be turned to port, and after changing its direction, be again formed into line, heading S., with the van squadron on the right, by the signals: *W.—head of fleet E.—S.*

23. The fleet being in line, heading N., to change front to the rear, turning to starboard, in reversed order, that is, with the van squadron on the left.

Fig. 52.

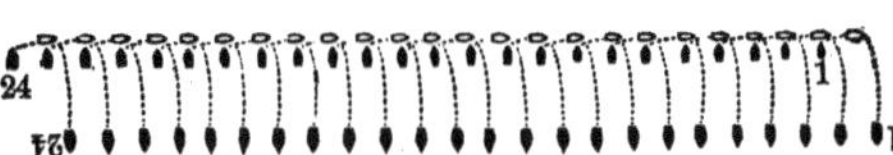

The commander-in-chief makes general signal E., which is repeated by the divisional commanders, when the fleet is formed into column of vessels heading E., from which it is thrown into line, heading S., upon general signal S.

It is evident that the fleet can be turned to port, and then formed into line, heading S., with the van squadron on the left, by the signals *W.* and *S.*

24. The fleet being in line, to form it into column to the rear.

FIG. 53.

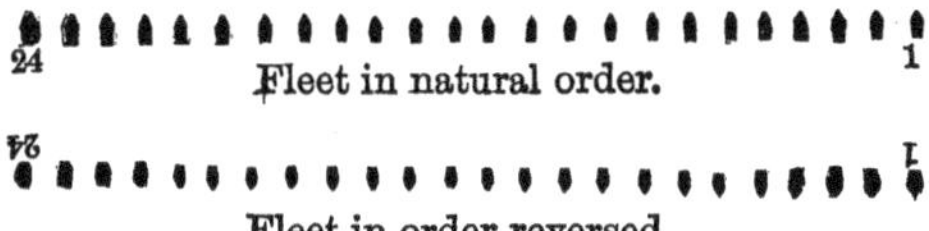

Fleet in natural order.

Fleet in order reversed.

The fleet should be first faced to the rear, and then formed into column as by the preceding methods, or it may be thrown into column to the right or left, at right-angles to the original direction, as in 10, Fig. 35; and then change direction to the course signalled by the commander-in-chief, as by 56, Fig. 87.

And it must be borne in mind that the right or left of a fleet in line is *its actual right or left on a line of bearing perpendicular to the course.* If, therefore, the signal be made, *From the right of fleet, form column of vessels,* number one will be the first to move forward if the fleet be in *natural order,* but number twenty-four if the fleet be in *reversed order.*

25. The fleet being in line, in close order, to form it in open order, on any vessel that may be designated.

To form it in open order on number twelve.

FIG. 54.

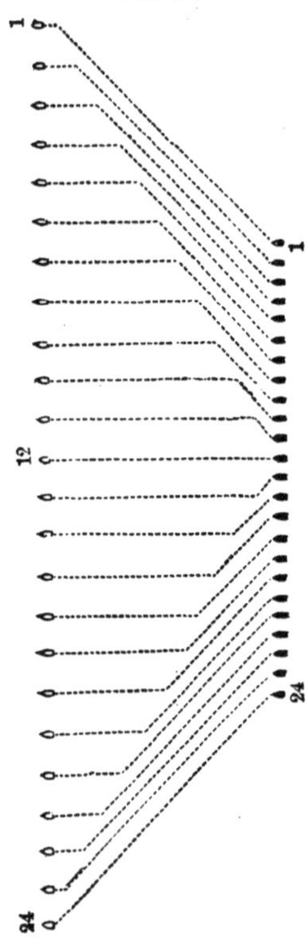

The commander-in-chief signals:

On the twelfth vessel—form in open order.*

Flag-ships of divisions signal: *On the twelfth* vessel—form in open order.*

Twelve continues her course under steerage-way; the vessels on her right steer four points to starboard, those on her left four points to port, under such steam as will enable them to preserve their line of bearing from her. When eleven and thirteen are in open order with respect to twelve, they resume their course and slow to steerage-way, at the same time hoisting the position pennant as a signal to ten and fourteen, which now form in open order upon them, and hoist the position pennant for the guidance of nine and fifteen, etc., etc. When all are in open order the fleet resumes its speed.

---

* Twelve's distinguishing pennant must here be hoisted.

26. The fleet being in line, in open order, to form it into close order, on any vessel which may be designated.

To form it into close order on number twelve.

FIG. 55.

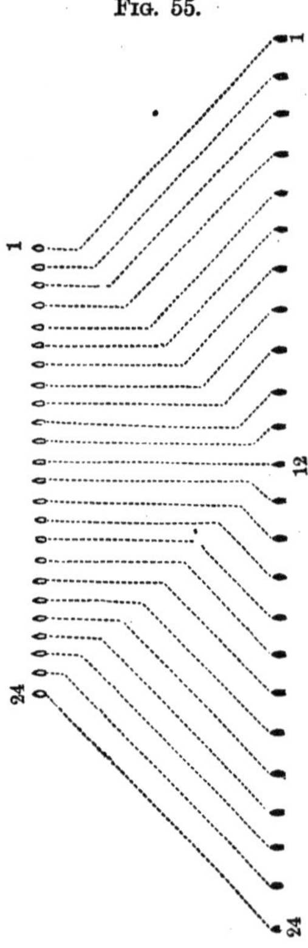

The commander-in-chief signals :

On the twelfth vessel—form in close order.

Flag-ships of divisions signal : *On the twelfth vessel—form in close order.*

Twelve continues her course under steerage-way; the vessels on her right keep four points to post, those on her left four points to starboard, under such steam as will enable them to preserve their line of bearing from her. When eleven and thirteen are in close order with respect to twelve, they resume their course and slow to steerage-way, at the same time hoisting the position pennant, as a signal to ten and fourteen, which now form in close order on them, and hoist the position pennant for the guidance of nine and fifteen, etc., etc.

When all are in close order, the fleet resumes its speed.

27. The fleet being in column of vessels, in natural order, to form it into double column.

Fig. 56.

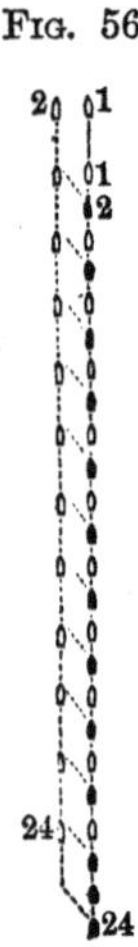

The commander-in-chief signals :

*Form double column—left oblique.**

Divisional commanders signal : *Division—form double column—left oblique.*

---

* The double column may be formed by a *right oblique* as well, in which case the even-numbered vessels will form to the right of the odd-numbered. Should the commander-in-chief then, however, wish to deploy it into line to the

The odd-numbered vessels continue onward under steerage-way; the others keep two points to port at full speed. When the latter find their consorts bearing from them on a line perpendicular to the course, they so regulate their speed as to maintain this bearing until they are at the proper distance from them, when they come to the course under steerage-way and hoist the position pennant. So soon as the double column is formed, the fleet resumes its speed.

28. The fleet being in columns of vessels, from the right of divisions to form it into double columns.

FIG. 57.

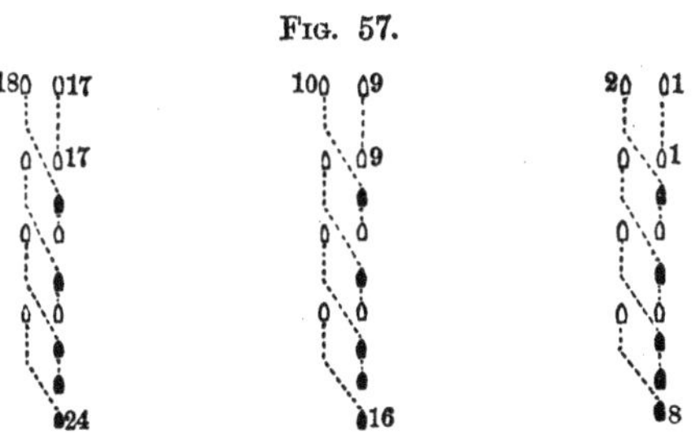

front, on the leading vessels, he will be forced to do it by a *right oblique*, and the line will be formed in order reversed—that is, with the van squadron on the *left*, whereas now he has but to form it by a *left oblique*, and it will be formed in natural order, as in Fig. 104.

The commander-in-chief signals :

*Form double column—left oblique.**

Flag-ships of divisions signal : *Division—form double column—left oblique.*

In each division, the odd-numbered vessels continue onward under steerage-way, the others keep two points to port at full speed. When the latter find their consorts bearing from them on a line perpendicular to the course, they so regulate their speed as to maintain this bearing until they are at the proper distance from them, when they come to the course under steerage-way, and hoist the position pennant.

So soon as the double columns are formed, the commander-in-chief signals the fleet to resume its speed.

---

* The *necessity* of forming the double columns by obliquing to the left is obvious here, for if formed to the right and then deployed into line to the front, eight would be on the right of the line and seventeen on the left, while sixteen would be *next to one* and *twenty-four next to nine* ; whereas now the commander-in-chief has but to signal : *Forward into line—left oblique,* and the line will be formed in natural order.

29. The fleet being in column of vessels, in natural order, to form it into triple column.

FIG. 58.

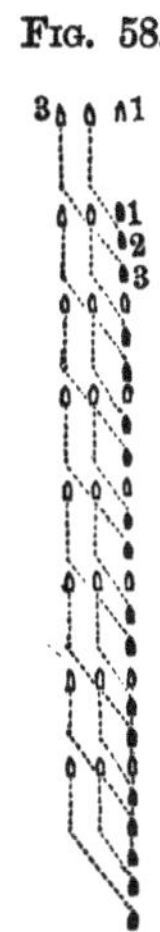

The commander-in-chief signals :

Form triple column—left oblique.

Flag-ships of divisions signal : *Division—form triple column—left oblique.*

One, four, seven, ten, thirteen, sixteen, nineteen, and twenty-two continue onward under steerage-way; the other vessels keep two points to port at full speed. When the latter find the former bearing from them on a line perpendicular to the course, they so regulate their speed as to maintain this bearing until they are at the proper distance from them, when they come to the course under steerage-way. To take the leading vessels as an illustration: When two finds one bearing from her on a line perpendicular to the course, she so regulates her speed as to maintain this bearing until she is at the proper distance from her, when she comes to the course under steerage-way, and hoists the position pennant for the guidance of three, which now manœuvres with respect to two, precisely as two has with respect to one.

When the triple column is formed, the commander-in-chief signals the fleet to resume its speed.

It is evident that the fleet can be formed into columns of fours, etc., etc., according to the same principles.

30. The fleet being in columns of vessels, from the right of divisions to form it into triple columns.

FIG. 59.

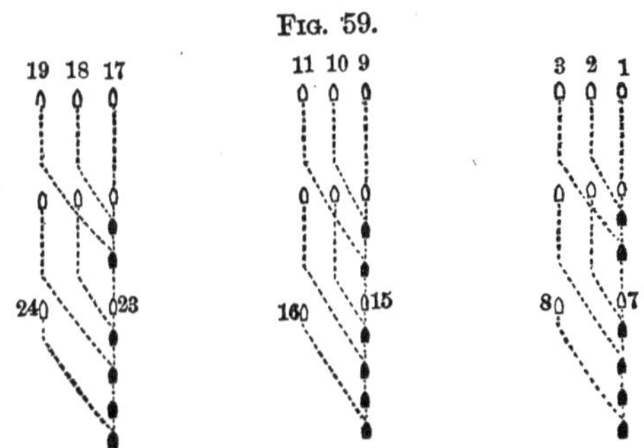

Commander-in-chief and divisional commanders signal, and the manœuvre is performed by each division, precisely as described in 29. There being but two vessels in each division to bring up the rear, they are disposed on the flanks.

It is evident that the divisions can be formed into columns of fours, etc., etc., accoding to the same principles.

31. The fleet being in double column, in natural order, to form it into triple column.

FIG. 60.

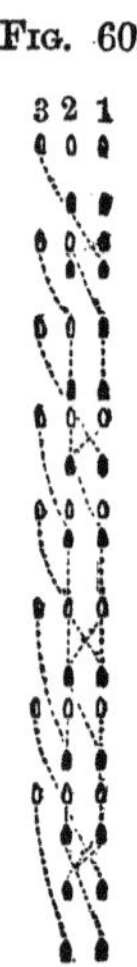

The commander-in-chief signals:

Form triple column—left oblique.

Flag-ships of divisions signal: *Division—form triple column—left oblique.*

One, two, seven, eight, nineteen and twenty continue onward under steerage-way ; four, ten, sixteen and twenty-two keep *four points* to starboard under steerage-way, and come into column in the wake of one, seven, thirteen, and nineteen respectively; the other vessels come two points to port at full speed, and manœuvreing get into position as in 29.

It is evident that the fleet, if steaming in double columns by divisions (as in Fig. 23), can be formed into triple columns by divisions, according to the same principles.

32. The fleet being in double column, in natural order, to form it into columns of fours.

Fig. 61.

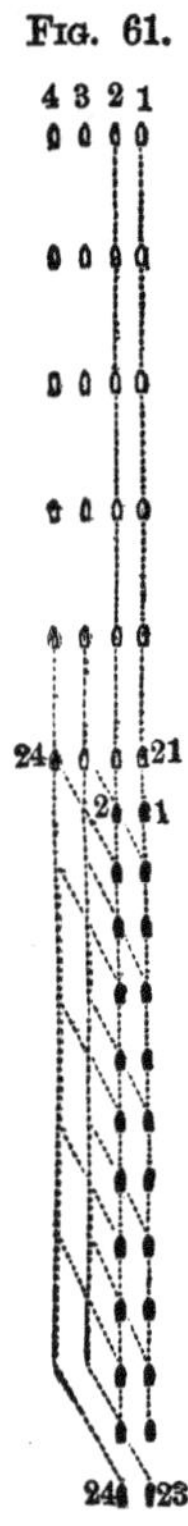

The commander-in-chief signals:

Form column of fours—left oblique.*

Flag-ships of divisions signal: *Division—form column of fours—left oblique.*

One, two, five, six, nine, ten, thirteen, fourteen, seventeen, eighteen, twenty-one, and twenty-two continue onward under steerage-way; the other vessels keep two points to port at full speed. When the latter find the former bearing from them on a line perpendicular to the course, they so regulate their speed as to maintain this bearing until they are at the proper distance from them, when they come to the course under steerage-way. To take the leading vessels as an illustration: When three finds one and two bearing from her on a line perpendicular to the course, she so regulates her speed as to maintain this bearing until she is at the proper distance from them, when she comes to the course under steerage-way, and hoists the position pennant for the guidance of four, which now forms on her.

It is evident that the fleet can be formed into column of sixes, eights,† etc., etc., according to the same principle.

---

* In this case, column of squadrons, and the signal may be as well: *Form column of squadrons.*

† In this case, column of divisions. (See Fig. 25.)

33. The fleet being in double columns, from the right of divisions to form it into columns of fours.

FIG. 62.

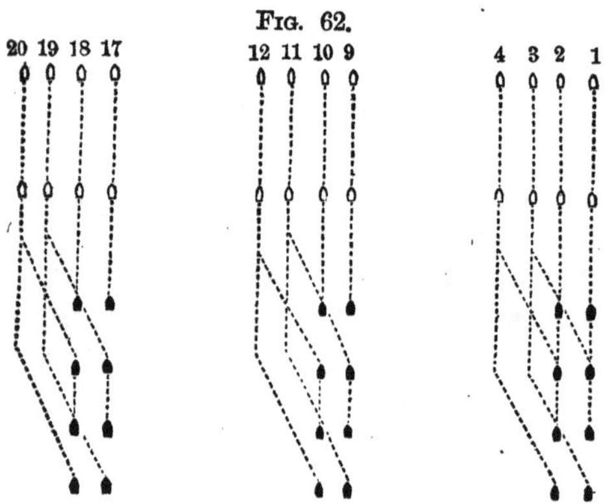

The commander-in-chief signals:

Form columns of fours—left oblique.

Flag-ship of divisions signal: *Division—form columns of fours—left oblique.*

Each division manœuvres precisely as in 32.

34. The fleet being in column of vessels, in natural order, heading N., to form it into columns of vessels abreast, by divisions, preserving the original direction.

To form it on the van division.

FIG. 63.

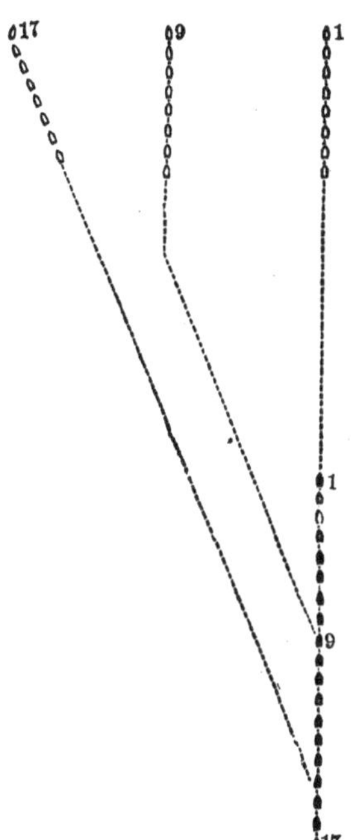

The commander-in-chief signals :

On the van division—form columns of vessels abreast by divisions, in natural order, heading N.

Flag-ship of van division signals · *Division N.*

Flag-ships of centre and rear divisions signal:* *Head of division N.N.W.*

The van division keeps its course under steerage-way; the leading ship of each of the other divisions obliques at once, at full speed to N.N.W.; the other vessels follow their leaders. When nine finds one bearing from her on a line perpendicular to the course (E. is the bearing in this case), she so regulates her speed as to maintain this bearing until she is at the proper distance from her, when she comes to the course under steerage-way and hoists the position pennant for the guidance of seventeen, which now manœuvres with respect to nine, precisely as nine has with respect to one. Of course, so soon as nine and seventeen slow, the vessels astern of them, respectively, slow.

---

* If the divisional commanders prefer it, they may make *general signal* to the vessels under their command *N.N.W.*, in which case all will keep N.N.W., being careful to preserve their bearing N. and S. from each other, so that when they come N. again, they will be in the *wake of their leaders.*

When the columns are formed the fleet resumes its speed.

It is evident that the fleet can be formed into columns of vessels abreast on the leading division, by a *right oblique* as well, but then the *van division would be on the left.*

To form it on the centre division.

FIG. 64.

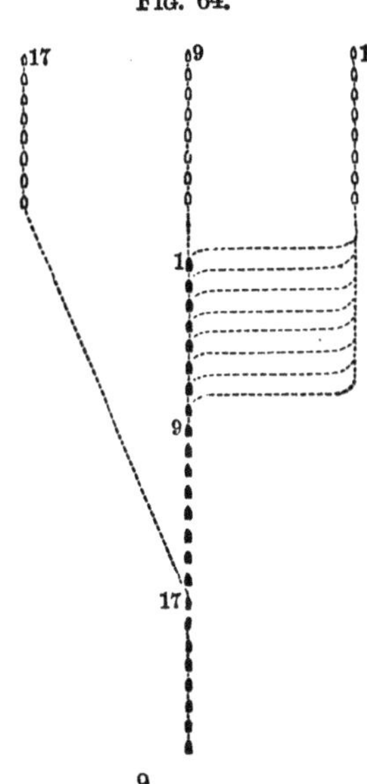

The commander-in-chief signals:

On the centre division—form columns of vessels abreast by divisions in natural order, heading N.

Flag-ship of van division : *Division E.**

Flag-ship of centre division : *Division N.*

Flag-ship of rear division: *Head of division N.N.W.*†

The van division comes E. and slows to steerage-way, until it has gained the proper distance from the centre division (continuing onward under steerage-way), when it comes N. by signal from the divisional commander. The leader of the rear division keeps N.N.W. at full speed, and is followed by the other vessels of this division, as they get into his wake. When seventeen finds nine bearing from her, on a line perpendicular to the course signalled by the commander-in-chief (E. is the bearing in this case), she so regulates her speed as to maintain this bearing un-

* Or the divisional commander may signal *head of division E.*, when the leading vessel (one) alone would come E., the other vessels standing on until in her *wake* before steering E., and again getting in her wake, after she has come N., before steering N. If the signal of the commander-in-chief were to form in reverse order, the van division would keep *W.*, and the rear one *N.N.E.*

† If the divisional commander prefers it, he may make *general signal* to the vessels under his command *N.N.W.*, in which case all will keep N.N.W., being careful to preserve their bearing N. and S. from each other, so that when they come N. again, they will be in the *wake of their leaders.*

til she is at the proper distance from her, when she comes to the course under steerage-way, and is followed by the other vessels of the division, successively, all of which have, of course, slowed to steerage-way when she slowed.

After the columns are formed the fleet resumes its speed.*

To form it on the rear division:

Fig. 65.

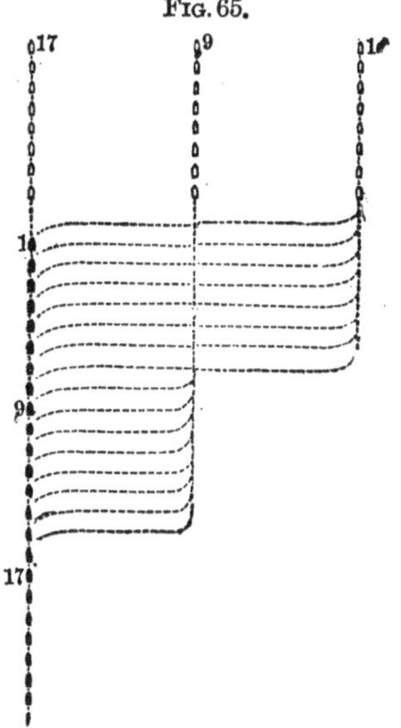

* Or so soon as nine finds seventeen abreast, she may go ahead full speed. See Figs. 63 and 64.

The commander-in-chief signals:

On the rear division—form columns of vessels abreast, by divisions, in natural order, heading N.

Flag-ships of van and centre divisions signal: *Division E.*

Flag-ship of rear division signals: *Division N.*

The van and centre divisions alter course together to E.; the rear division keeps its course. So soon as the centre division has gained its proper distance to the eastward of the rear one, it comes N., and hoists the position pennant for the guidance of the van division, which now manœuvres with respect to the centre division as the centre has with respect to the rear. The fleet can be formed by the van and centre divisions keeping W. as well, but will then be *in reversed order.*

To form the fleet into double columns, triple columns, etc., etc., abreast, by divisions, it should be first formed into columns of vessels abreast, by one of the above methods, and then into double columns, triple columns, etc., as in 28, 29, and 30.

It is evident that the fleet can be formed, by squadrons, into columns of vessels abreast, according to the same principles.

35. The fleet being in column of vessels, in natural order, heading N., to form it into columns of vessels abreast, by divisions, at right angles to the original direction.

To form it to the Eastward.

FIG. 66.

The commander-in-chief signals:

On the van division—form columns of vessels abreast, by divisions, in natural order, heading E.

Flag-ship of van division: *Head of division E.*

Flag-ships of centre and rear divisions signal: *Division N.*

The leading vessel of the van division comes E. at once, and is followed by the other vessels of the division, successively, as they get into her wake. So soon as the division is in column to the Eastward, it slows to steerage-way. The other divisions continue onward. When nine finds herself at the proper distance to the Northward of the van division she comes E., and is followed, successively, by the other vessels of the centre division as they attain her wake. So soon as the centre division is abreast of the van division, it slows to steerage-way, and hoists the position pennant for the guidance of the rear division,* which manœuvres with respect to the centre division, as the centre has with respect to the van. So soon as the columns are formed, the fleet resumes its speed. The fleet can be

---

* So soon as the leading vessel of this division has passed the rear of the centre division, steering E., it may *oblique* into position.

formed into columns of vessels abreast, heading E., by the leading vessel of each division coming E. at once, as shown by the dotted lines in the figure; but then it would be in order reversed.

It is evident that the fleet can be formed into columns of vessels abreast, heading W., according to the same principles. To form it in natural order then, however, the signal of commander-in-chief would be : *Heads of divisions W.*, whereas to form it in order reversed, he would have to signal : *On the van division—form columns of vessels abreast, by division—in reversed order—heading W.*

It is plain, too, that, in the same manner, it can be formed into columns of vessels abreast, by *squadrons*, at right angles to the original direction.

To form the fleet into double columns, triple columns, etc., etc., abreast, by divisions, at right angles to the original direction, it should first be formed into column of vessels abreast, by one of the above methods, and then into double columns, triple columns, etc., as in 29 and 30.

36. The fleet being in column of vessels heading N., in natural order, to form it into columns of vessels abreast, by divisions, on any course whatever.

To form it to the N.E.

Fig. 67.

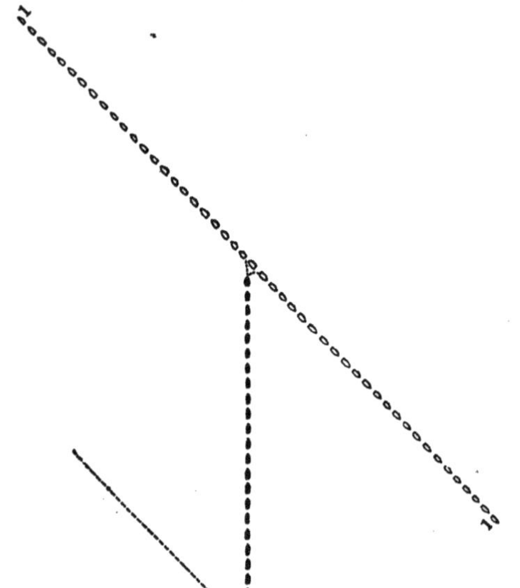

The fleet should be first formed on a line of bearing perpendicular to the given course, upon the signal, *Head of fleet S.E.*, or *head of fleet N. W.*, and afterwards into columns abreast, by divisions, on the given course, by the signal, *Heads of divisions N.E.**, as in manœuvre 35, Fig. 66.

It can also be by inverting the order of the fleet, so that the rear division may lead, heading S., and then proceeding as before.

It is evident that the fleet can be formed into columns of vessels abreast, by squadrons, on any point signalled, according to the same principles.

To form the fleet into double columns, triple columns, etc., abreast, by divisions or squadrons, on any course signalled, it should first be formed into columns of vessels abreast, by divisions or squadrons, as above, and then into double columns, triple columns, etc., etc., as in manœuvres 28, 29, and 30.

---

* Or he may make general signal N.E., and form the fleet into line, and afterwards proceed as in manœuvre 2, Figs. 19 and 20.

37. The fleet being in double column, to break it into column of vessels.

Fig. 68.

The commander-in-chief signals:

From the right, form column of vessels.

Divisional commanders signal: *Division—from the right, form column of vessels.*

One, three, five, seven, etc., etc., continue onward; two, four, six, eight, etc., etc., keep four points to starboard, so graduating their speed (see Table B) as to come into column at their proper distances, astern of one, three, five, seven, etc., etc., respectively.

The column can be formed inversely, according to the same principles, by breaking from the left.

38. The fleet being in triple column, to break it into column of vessels.

FIG. 69.

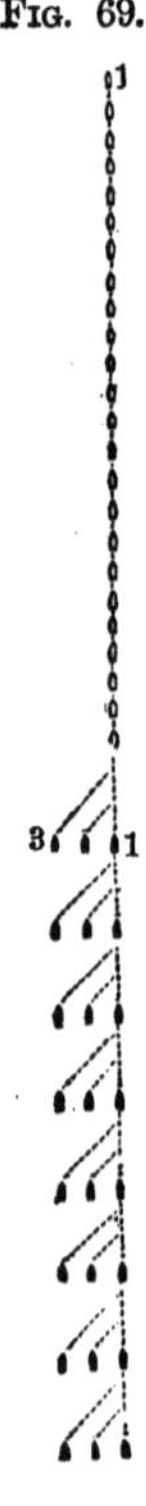

The commander-in-chief signals :

From the right, form column of vessels.

Divisional commanders signal : *Division—from the right, form column of vessels.*

Take the six leading vessels as an illustration.

One and four continue onward; two and three, and five and six, keep four points to starboard, so graduating their speed (see Table B) as to come into column in the wake of one and four respectively.

It is evident that the fleet may be broken into column of vessels from column of fours, column of fives, etc., etc., according to the same principles.

The column can be broken from the left, in the same manner, inversely.

39. The fleet being in double columns abreast, by divisions, to form it into columns of vessels abreast, by divisions.

FIG. 70.

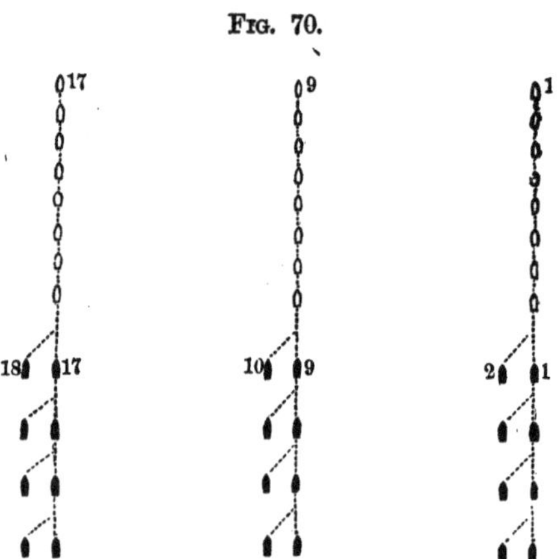

The commander-in-chief signals :

From the right of divisions, form column of vessels.

Divisional commanders signal: *Division—from the right, form column of vessels.*

Take the van division as an illustration:

One, three, five, and seven continue onward; two, four, six, and eight keep four points to starboard, so graduating their speed (see Table B) as to come into column at their proper distances astern of one, three, five, and seven respectively.

It is evident that a fleet steaming in columns abreast, by squadrons, could be broken into vessels abreast, by squadrons, according to the same principles.

It is also evident that if the fleet were in triple columns, columns of fours, etc., etc., abreast by divisions or squadrons, it could be broken into columns of vessels abreast, by divisions or squadrons, in the same way.

40. The fleet being in column of fours, to form it into double column.

Fig. 71.

The commander-in-chief signals:

From the right, form double column.

Divisional commanders signal: *Division—from the right, form double column.*

Take the eight leading vessels as an illustration: One and two, and five and six, continue onward; three and four, and seven and eight, keep four points to starboard, so graduating their speed (see Table B) as to come into column at their proper distances astern of one and two, and five and six, respectively.

It is evident that the fleet can be broken into double column, from column of sixes, column of eights, etc., according to the same principles.

41. The fleet being in column of fours abreast, by divisions, to form it into double columns abreast, by divisions.

Fig. 72.

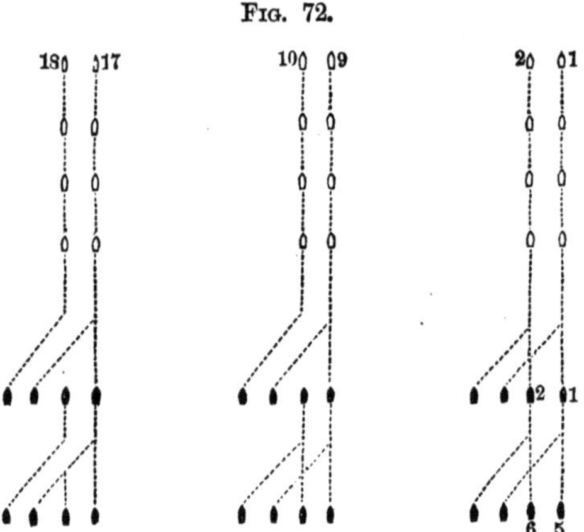

Commander-in-chief signals:

From the right of divisions, form double columns.

Divisional commanders signal: *Division—from the right, form double column.*

Take the van division as an illustration: One and two, five and six, continue their course; three and four, and seven and eight, keep four points to starboard under such steam as will enable them to come into column at their proper distances (see Table B) astern of one and two, and five and six, respectively.

42. The fleet being in columns of vessels abreast, by divisions, in natural order, heading N., to form it into column of vessels on the right division, in natural order, that is, with the van leading, and preserving the original direction.

Fig. 73.

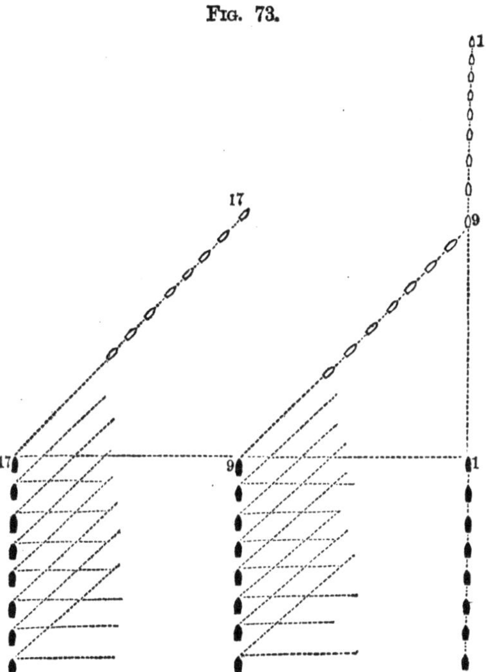

The commander-in-chief signals :

On the right division, form column of vessels—van in front—fleet N.E. (or E.)—right division N.

Flag-ship of van division signals: *Division N.*

Flag-ships of centre and rear divisions signal: *Head of division N.E. (or E.)**

The van division continues onward at full speed, the leading vessels of the centre and rear divisions keep N.E., while the other vessels stand on N. until in their leaders' wake, when they also come N.E.—all preserving such speed as is prescribed by Table B.† When the column is formed, the fleet proceeds at a uniform rate of speed.

It is evident that the fleet may be formed into column of vessels on the left division, with the rear (in left) in front, according to the same principles, *after reversing the fleet.* ‡

When the commander-in-chief wishes to leave it optional with the divisional commanders to

---

* Or they may make general signal to their divisions *N.E. or E.*, when all will oblique or wheel together, and go into column, as shown by the dotted lines.

† An inspection of Fig. 17, manœuvre 1, shows that this is analogous to the manœuvre for forming a fleet in line into column of vessels, preserving the original direction, where the speed of the obliquing vessels heading N.E. is seven knots, when that of the leading vessel (*division* or squadron in this case) is ten knots.

‡ For if the fleet were not reversed, the *rear squadron* would *not* be in front, although the rear division led.

come into column, either by *obliquing* or taking the two sides of the triangle, he omits signalling the *course to the fleet.*

43. The fleet being in columns of vessels abreast, by divisions, heading N., to form it into column of vessels on the centre division, in natural order; that is, with the van leading, and preserving the original direction.

FIG. 74.

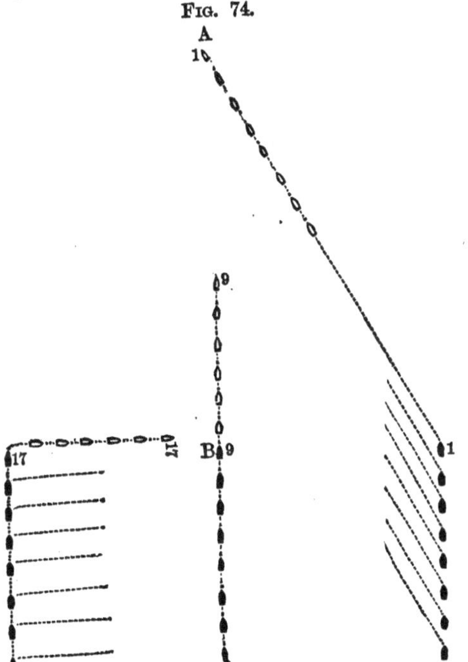

The commander-in-chief signals:

On the centre division, form column of vessels—van in front.

Flag-ship of van division: *Head of division N.N.W.* *

Flag-ship of centre division: *Division N.*

Flag-ship of rear division: *Head of division E.*†

The leading ship of the van division keeps N.N.W., and is followed by the other vessels of the division as they get into her wake, all moving at full speed. The centre division keeps onward at such speed ‡ as will enable it to come into

---

* Or he may make general signal to his division, *N.N.W.*, when all will keep N.N.W., as shown by the dotted lines.

† Or general signal to the division *E.*, and *move by the flank* into column.

‡ In close order, the distance between the divisions is 960 fathoms; it follows, then, that the leader of the van division steering N.N.W. will traverse 2,509 fathoms, and strike the perpendicular A B at a distance of 2,318 fathoms from B, the point of departure of the leader of the centre division, when the leaders should be 960 fathoms apart. Now, 960 taken from 2,318 leaves 1,358. Supposing, then, the van division to be steaming ten knots an hour, we have for the speed of the centre division $5\frac{4}{10}$ knots; for 2,509: 1,358: : 10: 5.4+. If the van be steaming fifteen knots, the speed of the centre must be $8\frac{1}{10}$ knots, for 10: 5.4+: : 15: 8.1+, etc., etc., etc. It is plain that the same reasoning will apply when all the vessels of the van division oblique together upon the signal *N.N.W.*, as is shown by the dotted lines.

If the divisions were in double column, triple column, etc., they could be formed into double column, triple

column at its proper distance (120 fathoms) from the rear vessel of the van division. The leading vessel of the rear division keeps E., and is followed by the other vessels of the division as they get into her wake, all moving at the same rate of speed as is maintained by the centre division.

So soon as the leader of the fleet finds herself bearing N. from the centre division, she hoists the *position* pennant, and comes to the course, *when all the fleet move at a uniform speed.**

It is evident that the column can be formed on the centre division, with the rear in front, according to the same principles.

---

column, etc., etc., on the van, rear, or centre division, according to the same principles; and if they are to be formed into column of vessels on the van, rear, or centre division, the divisions must be first broken into columns of vessels (see Fig. 70), and afterward manœuvre as in manœuvres 42 and 43.

A fleet in columns of vessels abreast, by squadrons, would be formed into column of vessels, on the van or rear, or right or left centre squadron, in the same manner.

* Should the commander of the centre division elect in this case, however, to move into column, as prescribed by Table J (that is, at the rate of $6\frac{7}{10}$ knots an hour, supposing the speed of the van division to be ten knots), the centre division would not steam ten knots until the *rear* vessel of the van division bore N. from it, as is plain; at which time the leader of the centre would have traversed 2318 fathoms and the leader of the van 2509+960 fathoms.

$$3469 : 2318 :: 10 : 6.7$$

44. The fleet being in columns of vessels abreast, by divisions, in natural order, heading N., to form it into column of vessels on the right division, with the van leading, on any course to the Northward, from E. to W.

To form the column to the N.E. or N.W.

FIG. 75.

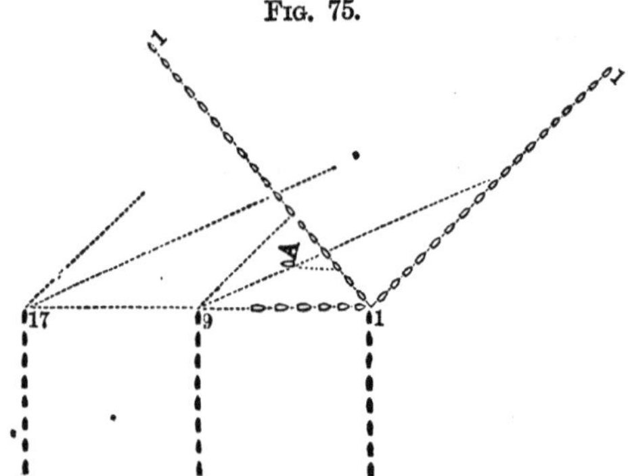

The commander-in-chief signals :

*On the right division, form column of vessels—van in front—head of van division N.E. (or N.W.)—heads of centre and rear divisions E.**

* Or he may signal an *oblique* course, when the divisions oblique into column, as shown by the dotted lines. Supposing them to steer E.N.E., to form the column to the N.E., and N. W. to get into column N.W., we have for their speed in the former case $9\frac{2}{10}$ knots, and 4.1 in

Flag-ship of van division: *Head of division N.E.* (*or N.W.*)

Flag-ships of centre and rear divisions signal: *Head of division E.**

The leader of the van division comes N.E. (or N.W.) at once; the leaders of the centre and rear divisions keep E., all moving at full speed. The other vessels stand on N., those of the van division keeping N.E. (or N.W.), those of the centre and rear divisions E., as they get into their leaders' wake.

When the divisions are in double columns, triple columns, etc., etc., they should be broken into columns of vessels, and then manœuvred as above.

When the commander-in-chief desires to form the column *toward the fleet*, at an angle of more than six points with the original course, he signals a course, making an angle of 45°, and when

---

the latter, provided the speed of the division on which the column is formed be ten knots. (See Tables F and L.) When the commander-in-chief wishes to leave the manner of forming column optional with the divisional commander, he omits signalling the course to the *centre and rear divisions.*

* Or they may make general signal *E.*, and move by the flank into column. In this case the signal of the commander-in-chief would be, *Centre and rear divisions E.—head of van division N.E.* (*or N.W.*)

he finds that the leader of the fleet has gone far enough to avoid collision, changes direction to the requisite course. Supposing the fleet to be steering N., for instance, and that the column is to be formed due W., the commander-in-chief first signals: *On the right division, form column of vessels—van in front—head of right division N. W.—heads of centre and rear divisions E.*, and afterward, *head of fleet W.*, and the leader of the fleet proceeds as shown by the vessel marked A.

A fleet in column of vessels abreast, by squadrons, would be similarly manœuvred.

45. The fleet being in column of vessels abreast, by divisions, in natural order, heading N., to form it into column of vessels, on the right division, with the van leading, on any course to the Southward from E. to W.

To form the column to the S.E. or S.W.

FIG. 76.

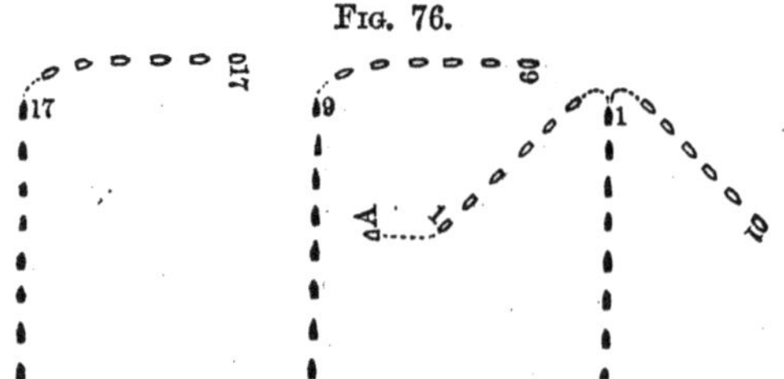

The commander-in-chief signals:

On the right division, form column of vessels—van in front—heads of centre and rear divisions E.—head of van division S.E. (*or S. W.*)

Flag-ship of van division signals: *Head of division S.E.* (*or S. W.*)

Flag-ships of centre and rear divisions signal: *Head of division E.*

The leader of the van division comes S.E. (or S.W.) at once; the leaders of the centre and rear divisions keep E., all moving at full speed. The other vessels stand on N., those of the van division keeping S.E. (or S.W.), those of the centre and rear divisions E., as they get into their leaders' wake. So soon as the commander of the centre division finds his leading vessel approaching the wake of the van division, steering S.E. (or S.W.), he signals, *Head* of division S.E. (or S.W.), etc., etc., as is apparent.

When the divisions are in double column, triple column, etc., etc., they should be broken into columns of vessels, and then manœuvred as above.

When the commander-in-chief desires to form the column to the rear, toward the fleet, he signals a four point course, and when he finds that the leader of the fleet has brought the rear vessels of the centre division to bear in the direction he wishes to steer, he changes direction to this course.

Supposing that in Fig. 76, the column is to be formed due W., the commander-in-chief should signal as above, and afterward, *Head of column W.*, when the leader would proceed as shown by the vessel marked A.

The fleet may be formed into column on the left division, with the rear in front, according to the same principles, on any course, *after reversing it*, as in 47, Fig. 78.

46. The fleet being in column of vessels, heading N., to change front to the rear, turning to starboard.

FIG. 77

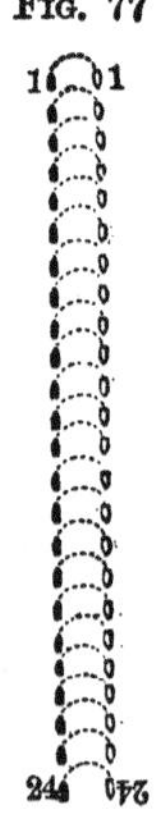

The commander-in-chief makes general signal *E.*, which is repeated by the divisional commanders, and as soon as he observes that the whole fleet is turning in that direction, signals *S.*, the divisional commanders again repeating his signal.

Each vessel, putting her helm to port, swings to starboard, until heading S.

It is evident that the fleet can change front to the rear, turning to port, according to the same principles.

47. The fleet being in column of vessels abreast, by divisions, heading N., to change front to the rear, turning to starboard.

Fig. 78.

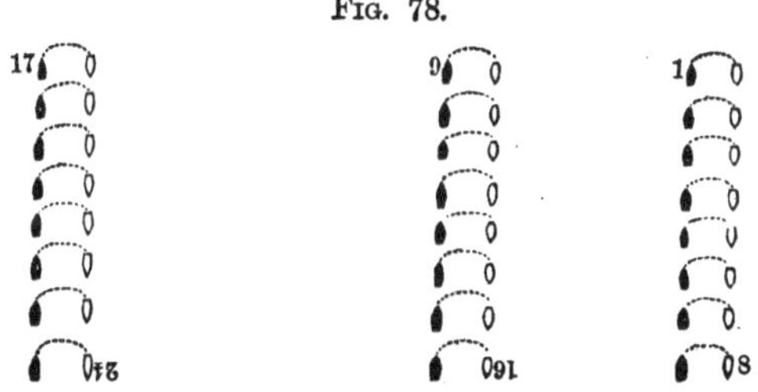

The commander-in-chief signals:

E., and afterward *S.*

Divisional commanders repeat these signals.

Performed by each division as in manœuvre

46. If the fleet were in columns of vessels abreast, by squadrons, front would be changed to the rear, according to the same principles.

48. The fleet being in double column, heading N., to change front to the rear, turning to starboard.

Fig. 79.

Signals made and manœuvre performed precisely as in manœuvre* 46.

It is evident that front could be changed to the

*If preferred, the fleet may be broken into column of vessels, and front changed to the rear, as in Fig. 77, and then re-formed into double column.

rear if the fleet were in triple column, column of fours, etc., etc. according to the same principles, by turning to port or starboard.

49. The fleet being in double columns abreast, by divisions, heading N., to change front to the rear, turning to starboard.

FIG. 80.

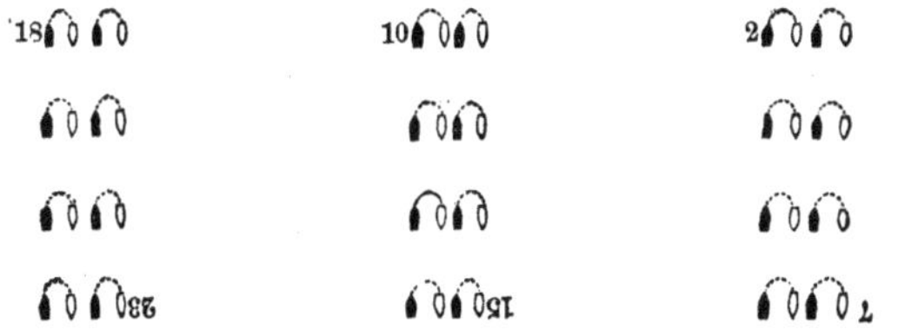

Commander-in-chief signals:

E., and afterward *S.*

Divisional commanders: *E.*, and afterward *S.*

Performed by each division as in 47. If the fleet were in triple columns, columns of fours, etc., etc., abreast, by divisions, or in double column, triple column, etc., abreast, by squadrons, front would be changed to the rear, according to the same principles.*

* See note to manœuvre 48.

50. The fleet being in column of vessels, heading N., to GAIN SEA to the left.

FIG. 81.

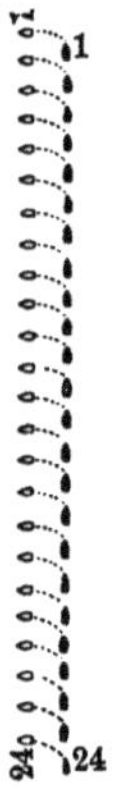

1 1

24 24

The commander-in-chief signals:

W.

Divisional commanders signal: *Division W.*

Each vessel putting her helm to starboard, swings to port until on the required course; the fleet (now in line) continues on this course until it has gained the proper distance, and the commander-in-chief makes general signal *N.*, when it comes N., and is in column of vessels again.

It is evident that sea can be gained to the right, according to the same principles.

51. The fleet being in columns of vessels abreast, by divisions, heading N., to gain sea to the left.

Fig. 82.

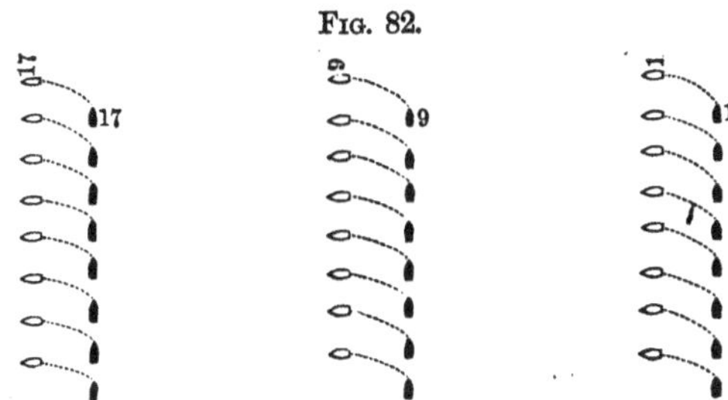

The commander-in-chief signals :

W.

Divisional commanders signal: *Division W.*

Each division comes W.; the fleet (now in column of divisions with the *rear or left in front*) continues on this course until the commander-in-chief signals N., when it resumes its original course and formation.

It is evident that sea may be gained to the right, according to the same principles.

It is also evident that a fleet in columns of vessels abreast, by squadrons, to gain sea to the right or left would be similarly manœuvred.

52. The fleet being in double column, heading N., to gain sea to the left.

FIG. 83.

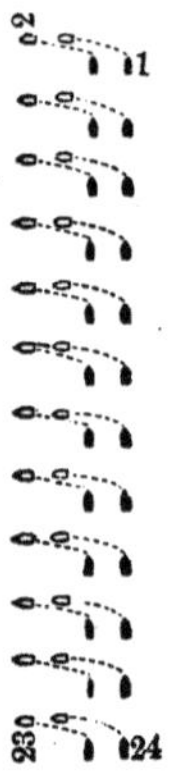

The commander-in-chief signals:

W.

Divisional commanders signal: *Division W.*

All come W., and the fleet moves by the flank in this direction, until the commander-in-chief signals N., when it resumes its original direction and formation.

It is evident that a fleet in triple column, column of fours, etc., etc., would gain sea to the left or right, according to the same principles.

53. The fleet being in double columns abreast, by divisions, to gain sea to the left.

FIG. 84.

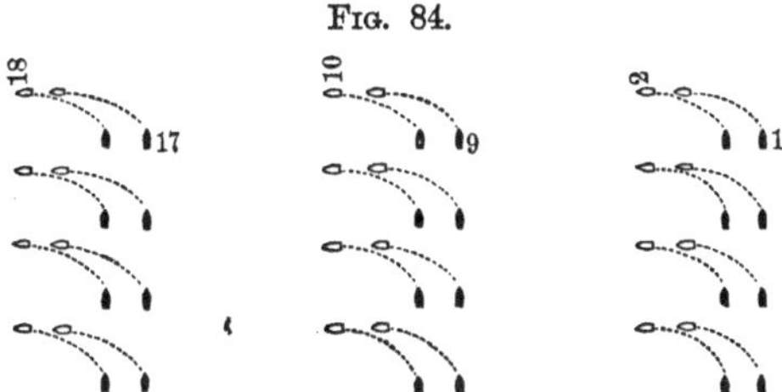

The commander-in-chief signals:

W.

Divisional commanders signal: *Division W.*

All come W., and the divisions move by the flank, until the commander-in-chief signals *N.*, when they resume their original direction and formation.

It is evident that if the fleet were in double columns abreast, by squadrons, it would be similarly manœuvred, and if in triple columns, columns of fours, etc., etc., abreast, by divisions or squadrons, it would gain sea to the right or left, according to the same principles.

54. The fleet being in column of vessels, heading N., to gain sea to the front and left.

Fig. 85.

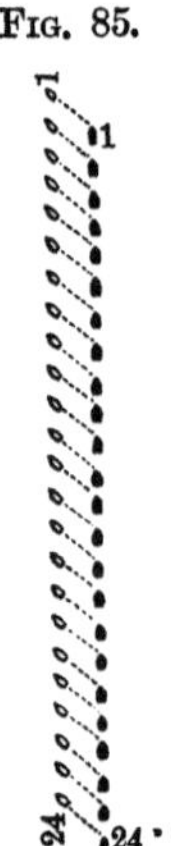

The commander-in-chief signals :

*N. W.**

* If the commander-in-chief desires to move more to the front, he signals further to the N., N. by W., or N.N.W. for instance; if more to the left, further to the W.

Each vessel, putting her helm to starboard, swings to port until heading N.W., and the fleet moves by the flank,* until the commander-in-chief signals *N.*, when it resumes its original direction and formation.

Sea can be gained to the right and front, according to the same principles; and if in double column, triple column, etc., etc., the fleet would be similarly manœuvred.

55. The fleet being in column of vessels abreast, by divisions, heading N., to gain sea to the front and left.

FIG. 86.

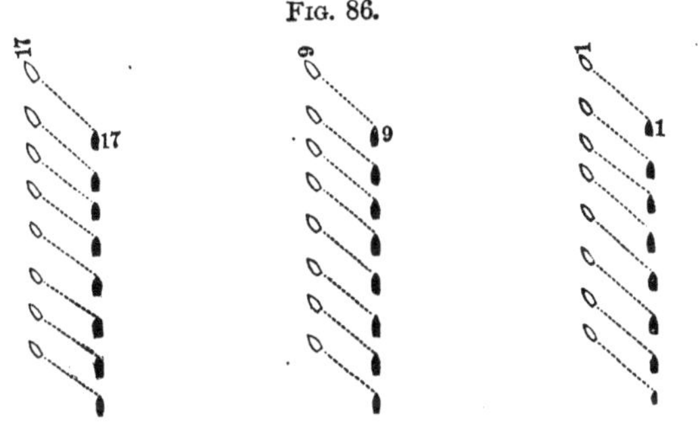

* Great care must be used by the vessels to maintain their line of bearing N. and S., so that when they resume their original direction they may be in the wake of their leader (or leaders).

12

The commander-in-chief and divisional commanders signal *N. W.*, and each division moves by the flank until the commander-in-chief signals *N.*, when the fleet resumes its original course and formation. If the divisions were in double column, triple column, column of fours, etc., etc., they would be similarly manœuvred.

It is evident that sea can be gained to the front and right, according to the same principles.

See notes to 54.

56. The fleet being in column of vessels, heading N., to change direction to the left.

Suppose, for instance, it be required to steer N.W.

FIG. 87.

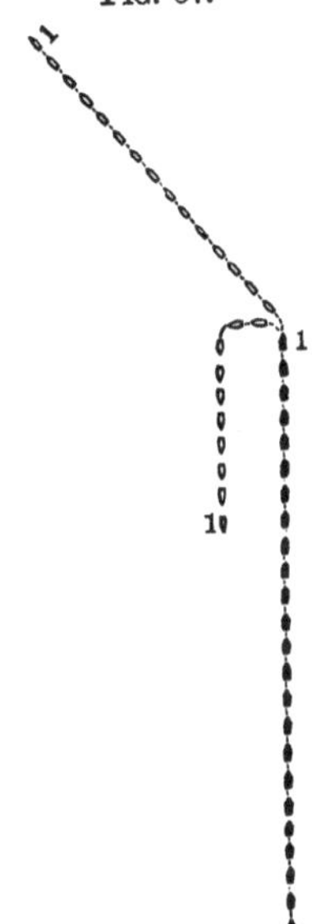

The commander-in-chief signals:

Head of fleet N.W.

Flag-ships of leading division : *Head of division N. W.*

Flag-ships of centre and rear divisions* signal: *Division N.*

The leader of the fleet alters course at once to N.W.; the other vessels continue onward, two steering N.W. when in the wake of one, three when in the wake of two (after this vessel has come to the required course), and so on, to the last vessel.

When the angle of the new course with the old is over ninety degrees, each vessel must be careful to turn where her next ahead has turned, and to follow in her wake.

Where the commander-in-chief wishes to change the direction of the fleet to a directly opposite one, turning to port—to S. in this example—turning to the Westward, he has but to make signal, *Head of fleet W.*, and afterward, as he sees the leader turning to port, *Head of fleet S.*

It is evident that direction may be changed to the right, according to the same principles.

* Or these commanders may omit making signals until the heads of their divisions are approaching the rear of the divisions preceding them, *steering N. W.*, when they must of course signal : *Head of division N. W.*

57. The fleet being in columns of vessels abreast, by divisions, in natural order, heading N., to change direction to the left, on any course from N. to W.

Suppose, for example, it be required to steer N.W.

1*st Method.*

FIG. 88.

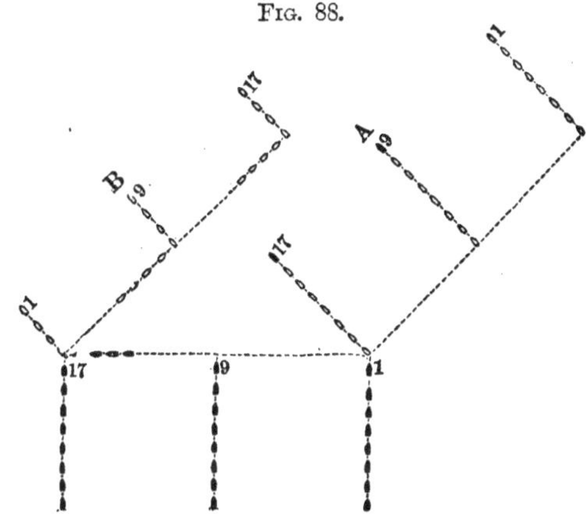

The commander-in-chief signals :

On the right division, form column of vessels—van in front—head of centre and rear divisions E.—head of van division N.E.*

* If the commander-in-chief wishes the divisional commanders to choose their own way of going into column, he omits signalling the course to the *centre* and *rear* divisions.

The flag-ship of van division signals: *Head of division N.E.*

Flag-ships of centre and rear divisions signal: *Head of division E.*

Performed as in manœuvre 44, Fig. 75, and so soon as the commander-in-chief observes that the leader of the rear division *is approaching the wake of the van and centre divisions, steering N.E.*, he signals: *Heads of divisions N.W.;* and the manœuvre is completed, as shown by A.

It is evident that direction may be changed to the left, by forming the column to the N.E., on the rear division; but then the fleet will be in order reversed, the van division being on the left, as shown by B.

If the divisions are in double column, triple column, etc., they must be broken into columns of vessels, and re-formed into any order required, after direction has been changed.

2d *Method.*

FIG. 89.

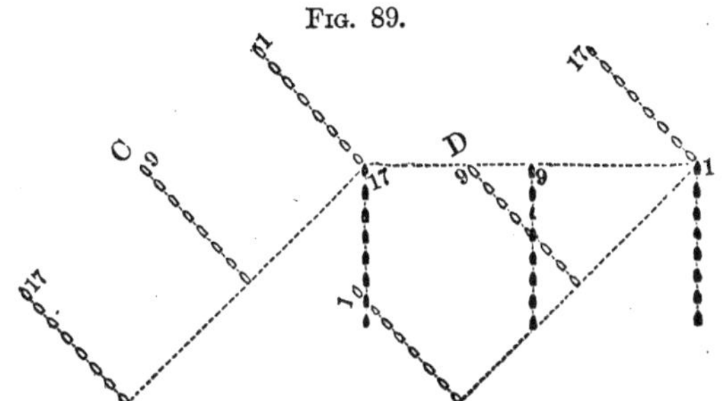

The commander-in-chief signals :

On the left division, form column of vessels—rear in front—head of van and centre divisions W.—head of rear division S. W.*

The flag-ships of van and centre divisions signal: *Head of division W.*

Flag-ship of rear division signals: *Head of division S. W.;* and the manœuvre is performed as explained in the 1st Method, and shown by C.

It is evident that direction may be changed to the left by forming the column to the S.W., on the van division; but then the fleet will be in order reversed, the van division being on the left, as shown by D.

---

* If the commander-in-chief please, he may omit signalling the course to the *van* and *centre* divisions. (See note to 1st Method.)

58. The fleet being in columns of vessels abreast, by divisions, in natural order, heading N., to change direction to the right, on any course from N. to E.

Suppose, for example, it be required to steer N.E.

1*st Method.*

FIG. 90.

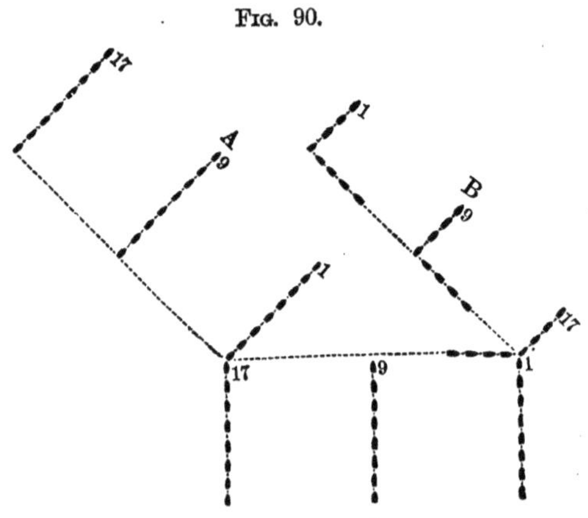

The commander-in-chief signals :

*On the left division, form column of vessels—rear in front—head of van and centre divisions W.**—*head of rear division N. W.*

* See note to manœuvre 57.

The flag-ships of van and centre divisions signal: *Head of division W.*

The flag-ship of rear division signals: *Head of division N. W.*

Performed as explained in manœuvre 44, and shown by Fig. 90, and so soon as the commander-in-chief observes that the leader of the van division is approaching the wake of the rear and centre divisions, steering N.W., he signals: *Heads of divisions N.E.*, and the manœuvre is completed, as shown by A.

It is evident that direction may be changed to the right, by forming the column to the N.W. on the van division; but then the fleet will be in order reversed, the van division being on the left, as shown by B.

2d Method.

FIG. 91.

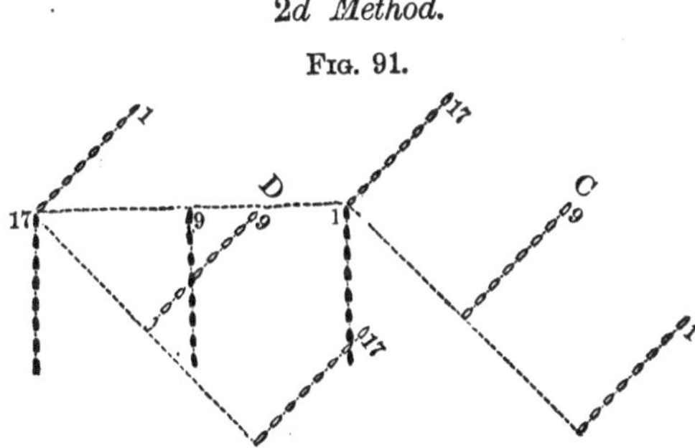

The commander-in-chief signals :

On the right division, form column of vessels—van in front—head of centre and rear divisions E.—head of van division S.E.

Flag-ship of van division : *Head of division S.E.*

Flag-ships of centre and rear divisions signal: *Head of division E.*

Performed as explained in manœuvre 45, Fig. 76 ; and so soon as the commander-in-chief observes that the leader of the rear division is approaching the wake of the van and centre divisions, steering S.E., he signals : *Heads of divisions N.E.*, and the manœuvre is completed, as shown by C.

Direction may be changed to the right, by forming the column to the S.E., on the rear division; but the fleet will be in order reversed, as shown by D.

59. The fleet being in columns of vessels abreast, by divisions, heading N., to change direction to the rear, on any course from S. to W.

Suppose it be required to steer S.W.

1*st Method.*

Fig. 92.

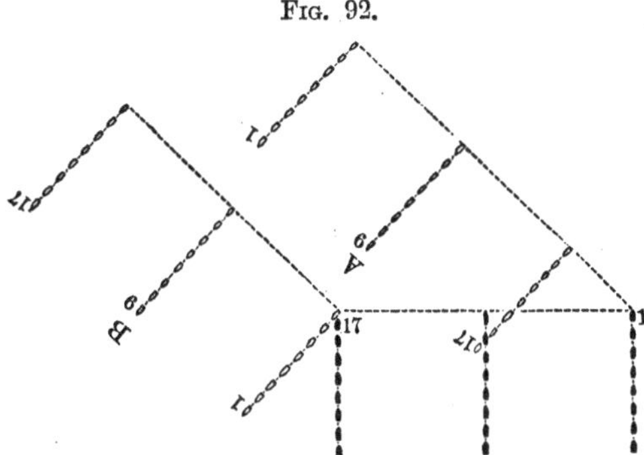

The commander-in-chief signals:

On the right division, form column of vessels—van in front—head of centre and rear divisions E.—head of van division N. W.

Flag-ship of van division: *Head of division N. W.*

Flag-ships of centre and rear divisions signal: *Head of division E.*

Performed as explained in manœuvre 44, Fig. 75, and then re-formed into columns abreast, heading S.W., upon signal from the commander-in-chief of, *Heads of divisions S.W.*, as shown by A.*

The column may be formed on the rear division to the N.W.; but the fleet will be in order reversed, as shown by B.

2d Method.

FIG. 93.

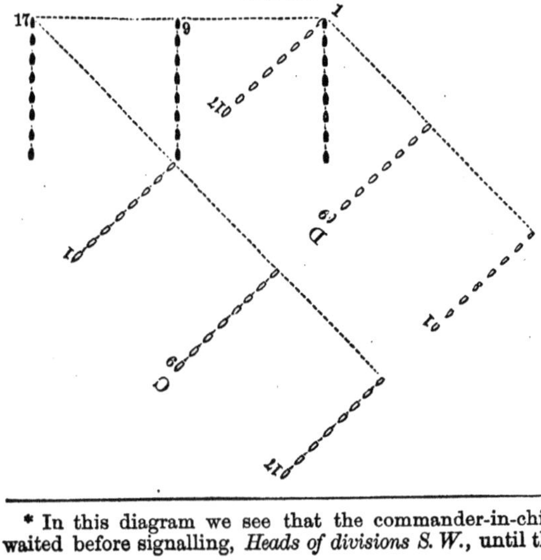

* In this diagram we see that the commander-in-chief waited before signalling, *Heads of divisions S.W.*, until the

The commander-in-chief signals :

On the left division, form column of vessels—rear in front—heads of centre and van divisions W.—head of rear division S.E.

Flag-ships of van and centre divisions signal : *Head of division W.*

Flag-ship of rear division signals : *Head of division S.E.*

Performed as explained in the 1st Method and shown by Fig. 93, and then re-formed into columns abreast, heading S.W., upon signal from the commander-in-chief of, *Heads of divisions S.W.*, as shown by C.*

The column may be formed to the S.E., on the van division ; but the fleet upon being reformed into columns of vessels abreast, heading S.W., will be in order reversed, as shown by D.

---

first five or six vessels of the rear division were in column after the van and centre divisions, steering N.W.

* In this diagram the commander-in-chief has formed the whole fleet into column of vessels to the S.E., before signalling, *Heads of divisions S. W.*

60. The fleet being in columns of vessels abreast, by divisions, heading N, to change direction to the rear, on any course from S. to E.

Suppose it be required to steer S.E.

1st *Method.*

FIG. 94.

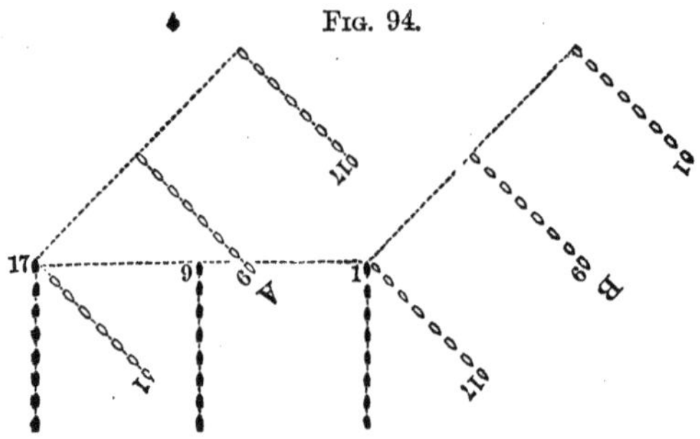

The commander-in-chief signals :

On the left division, form column of vessels—rear in front—heads of centre and rear divisions W.—head of rear division N.E.

Flag-ships of van and centre divisions signal : *Head of division W.*

Flag-ship of rear division signals : *Head of division N.E.*

Performed as explained in 57, and then re-formed into columns of vessels abreast, heading S.E., upon the signal, *Heads of divisions S.E.*, as shown by A.

The column may be formed to the N.E., on the van division; but then the fleet upon being re-formed into columns of vessels abreast, heading S.E., will be in order reversed, as shown by B.

2d Method.

FIG. 95.

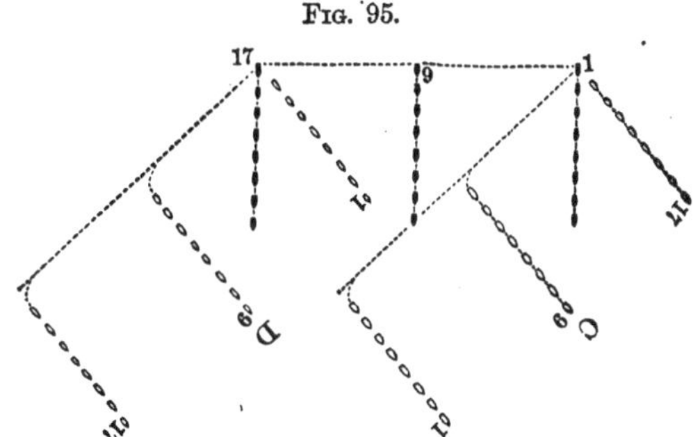

The commander-in-chief signals:

On the right division, form column of vessels—van in front—heads of centre and rear divisions E.—head of van division S. W.

Flag-ship of van division signals: *Head of division S. W.*

Flag-ships of centre and rear divisions signal: *Head of division E.*

Performed as explained in manœuvres 58 and 59; and so soon as the commander-in-chief observes that the leader of the rear division is approaching the wake of the van and centre divisions, steering S.W., he makes the signal: *Heads of divisions S.E.*, and the manœuvre is completed, as shown by C.

The column may be formed to the S.W., with the rear division leading; but then the fleet, upon being formed into columns abreast, by divisions, will be in order reversed, as shown by D.

61. The fleet being in column of vessels, in close order, to form it in open order on any vessel which may be designated.

To form it on the leading vessel.

FIG. 96.

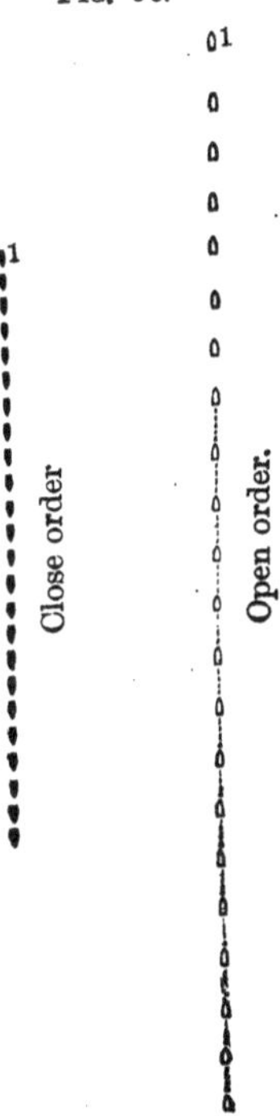

The commander-in-chief makes signal:

On the van vessel, form in open order.*

Which is repeated by the divisional commanders.

The leader of the fleet continues onward at full speed; all the other vessels slow to steerage-way, two resuming her speed, and hoisting the position pennant, when one has gained the proper distance from her, three when two has gained her proper distance, and so on to the last vessel.

To form in open order on the rear vessel, the commander-in-chief makes signal:

On the rear vessel, form in open order.

The rear vessel slows to steerage-way; all the other vessels continue onward at full speed, twenty-three slowing to steerage-way, and hoisting the position pennant when she has gained the proper distance from the rear vessel, twenty-two doing the same when she has gained her proper distance, and so with the rest of the fleet.

When the commander-in-chief signals one of the intervening vessels † as the one to form on,

---

* The distinguishing pennant of the vessel upon which the fleet is to form, is, of course, to be hoisted by the commander-in-chief.

† It is best to form from close to open order, and from open to close order on the van or rear vessel. The signal *close order* or *open order*, without the designation of any particular vessel, is always to be understood as an order to close or open on the *leader of the column*, whether the van or rear be in front.

the vessel designated (say No. 12 in this case) slows to half-speed, and the vessels astern of her to steerage-way, while the rest of the fleet keep at full speed. When eleven and thirteen find themselves in open order with respect to twelve, they regulate their speed by this vessel's, and hoist the position pennant as a signal to ten and fourteen, which now form in open order upon them, and hoist the position pennant as a signal to nine and fifteen, etc., etc., etc.

When all are in open order, the fleet resumes its speed. The fleet may be formed in open order or close order from half-distance, in a similar manner

62. The fleet being in column of vessels, in open order, to form it into close order, on any vessel that may be designated.

To form it on the leading vessel.

(See Fig. 96.)

The commander-in-chief makes signal :

On the van vessel, form in close order.

Which is repeated by the divisional commanders.

The leader of the fleet slows to steerage-way; all the other vessels continue onward at full speed, two slowing to steerage-way, and hoisting the position pennant as a guide to three, when she is in position, and so on, in succession, to the last vessel.

When the column is to close on the rear vessel, this vessel alone preserves her speed, the other vessels slowing to steerage-way. When twenty-three finds that the rear vessel is about closing up to close order, she resumes her speed, and hoists the position pennant as a signal to the next ahead, and so on to the van vessel.

To close upon one of the intervening vessels, number twelve, for instance :*

Twelve steams at half-speed, the vessels astern of her at full speed, and the vessels ahead under steerage-way. When eleven and thirteen find themselves in close order with respect to twelve, they regulate their speed by that vessel's, and hoist the position pennant as a signal to ten and fourteen, which in turn, when in position, hoist the position pennant as a signal to nine and fifteen, etc., etc., etc.

When all are in close order, the fleet moves at a uniform rate of speed.

It is evident that the fleet could be closed, to half distance, from open or close order, in a similar manner.

---

* See note to 61.

63. **The fleet being in double column, in close order. to form it in open order upon any vessel that may be designated.**

Fig. 97.

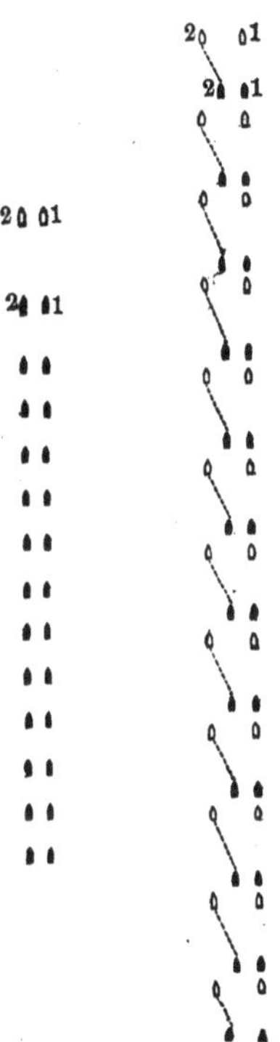

To form it on the leading vessel.

The commander-in-chief signals:

On the van vessel, form in open order.*

Divisional commanders signal: *On the van vessel, form in open order.*

The leading vessels continue onward at full speed; the other vessels slow to steerage-way; three and four resuming their speed and hoisting the position pennant when one and two have gained their proper distance from them, five and six doing the same when three and four have gained *their* distance; and so on to the closing vessels of the column.

The *odd-numbered* vessels now slow, while *even-numbered* vessels keep two points to port, at such speed as will enable them to keep their consorts on a line of bearing perpendicular to the course, until they are distant from them two cables' length; when they resume their original direction, and the fleet resumes its speed.

It is evident that the column may be formed on the rear or centre vessels, as explained in 61.

If the fleet were in triple column, column of

---

* Number one's distinguishing pennant is here hoisted. If two's pennant were hoisted, instead, the *odd-numbered* vessels would keep two points to starboard to open the columns.

fours, etc., etc., in close order, it would be formed in open order, according to the same principles.

It is evident that a fleet in double column, at half distance, may be formed in close or open order, in a similar manner.

64. The fleet being in double column, in open order, to form it in close order, upon any vessel that may be designated.

FIG. 8.

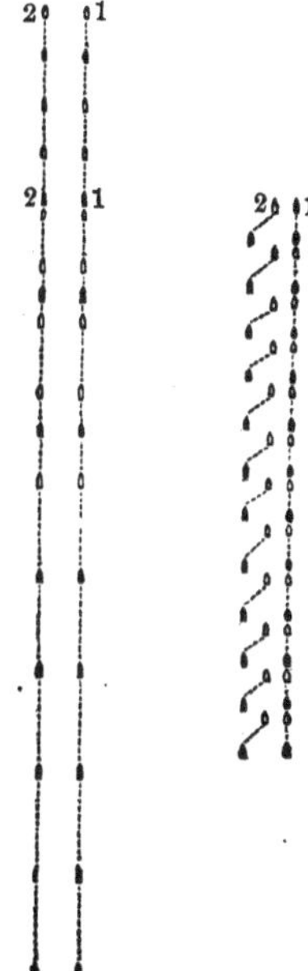

To form it on the leading vessels.

The commander-in-chief signals:

On the van vessel, form in close order.

Divisional commanders signal: *On the van vessel, form in close order.*

The leading vessels slow to steerage-way; the others continue onward at full speed, three and four slowing to steerage-way, and hoisting the position pennant when they have closed up, etc., etc., etc.

The even-numbered vessels now keep two points to starboard, at such speed as will enable them to keep their consorts on a line of bearing perpendicular to the course, until they have closed to a cable's distance from them; when they resume their original direction, and the fleet resumes its speed.

It is evident that the column may be formed on the rear or centre vessels, inversely, as explained in manœuvre 60.

If the fleet were in triple column, column of fours, etc., etc., in open order, it would be formed in close order, according to the same principles.

The fleet would also be formed at half-distance from open or close order, according to the same principles.

65. The fleet being in columns of vessels abreast, by divisions, in close order, to form it in open order.

FIG. 99.

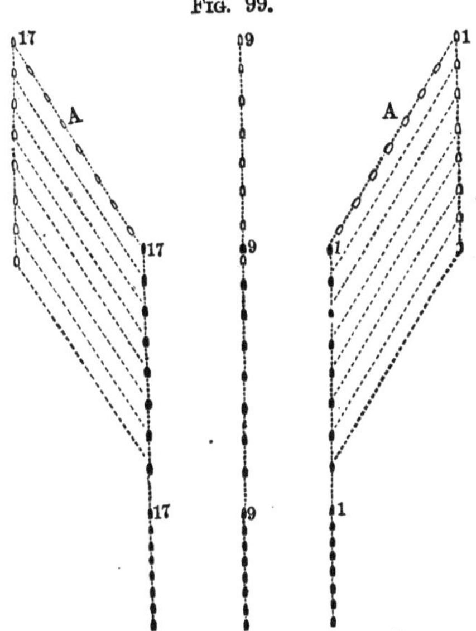

The commander-in-chief makes signal:

On the van vessels, form in open order; or, *On the rear vessels, form in open order.*

Which signal is repeated by the divisional commanders.

Performed by each division as explained in 61,

the vessels upon which the formation is made being careful to preserve their line of bearing abreast during the manœuvre.

Should the commander-in-chief not make further signal, the divisions will now proceed onward at a uniform rate of speed, the vessels of each division being in open order, while the divisions are with respect to each other in close order.

Suppose, however, he desires to form the divisions in open order on the centre division. The signal will be, *On the centre division, form in open order.*

The centre division slows to steerage-way; the van and rear divisions keep two points to starboard and port, respectively,* at such speed as will enable them to keep on a line of bearing, abreast of the centre division. So soon as one and seventeen find they have gained the proper distance from nine, they hoist the position pennant and come to the course, followed by all the vessels of their divisions. The centre division now resumes its speed.

To form on the van division, the centre and

---

* If the commanding officers of the van and rear divisions think proper, they may signal, *Head of division N. N. E.—head of division N. N. W.* (supposing the fleet to be heading N.), and gain distance as shown by *a a*.

rear divisions keep two points to port, while the van maintains its course.

To form on the rear division, the van and centre divisions keep two points to starboard, and the rear division preserves its course.

It is evident that the fleet may be formed in close order from open order, or at half distance from close or open order, according to the same principles, inversely.

66. The fleet being in column of vessels, heading N., to form it into line to the front.

1*st Method.*

FIG. 100.

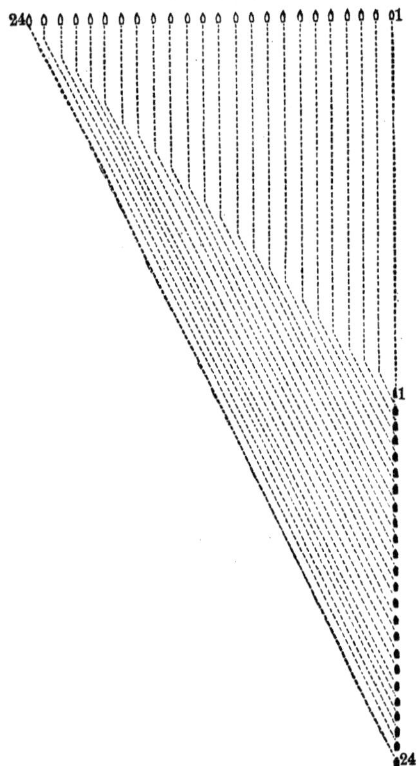

The commander-in-chief makes signal:

Forward into line—left oblique.

Flag-ship of van division: *Division N. N. W.—van vessel N.**

Flag-ships of centre and rear divisions signal: *Division N. N. W.*

The leading vessel continues onward under steerage-way; the other vessels keep two points to port at full speed. When two finds one bearing from her on a line perpendicular to the course (E. is the bearing in this case), she regulates her speed so as to maintain the bearing until she is at the proper distance from one, when she hoists the position pennant, and comes to the course under steerage-way. Three manœuvres similarly with respect to two, four with respect to three, *and so on to the last vessel.*

The line may be formed by obliquing to the right; but the vessels will then be in order reversed, the van vessel being on the left.

* Vessel's distinguishing pennant will, of course, be shown; so if the left (or rear) were in front, steering N., twenty-four's pennant would be displayed, and the centre and van divisions would signal, *Division N. N. E.*

2d Method.

FIG. 101.

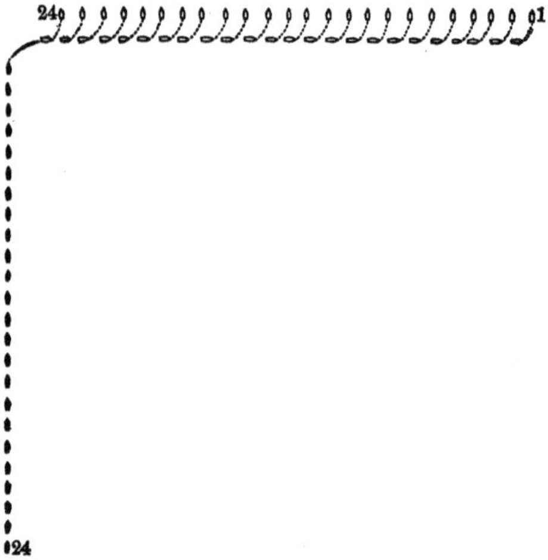

24 ... 1

24

The commander-in-chief signals:

Head of fleet E.

Flag-ship of* van division: *Head of division E.*

Flag-ships of centre and rear divisions signal: *Division N.*†

---

* See note to 1st Method.

† Or omit signalling here, being careful to signal when they observe the leading vessels of their divisions approaching the wake of the van division steering E., *Head of division E.*

The leader of the fleet keeps E. at once, the other vessels stand on N. and come E. in succession in her wake. So soon as the column is formed to the Eastward, the commander-in-chief makes general signal *N.*, and the manœuvre is completed.

The manœuvre may be performed by forming the column to the Westward; but the fleet on coming N. will then be in order reversed.

67. The fleet being in columns of vessels abreast, by divisions, heading N., to form it into line to the front.

1st Method.

Fig. 102.

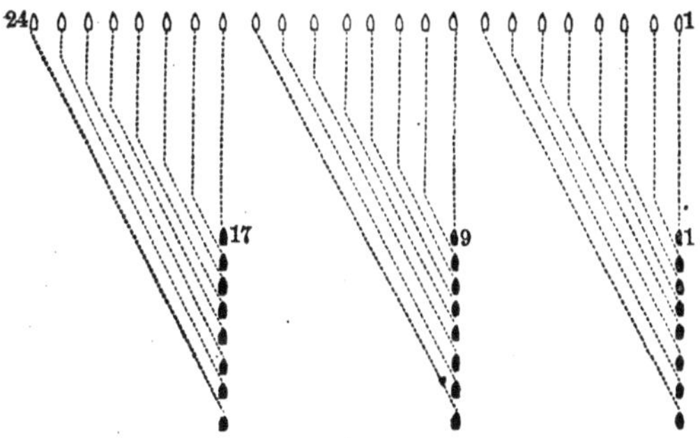

The commander-in-chief signals :

Forward into line—left oblique.

Divisional commanders signal: *Division N.N.W. —leading vessel N.*

Performed by each division, as explained in 66, 1st Method.

2d Method.

FIG. 103.

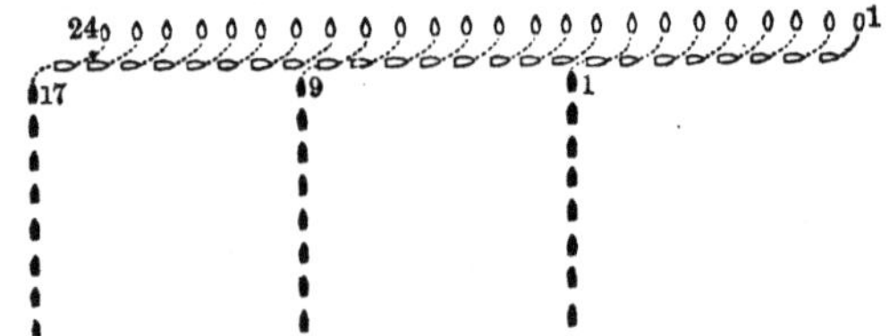

The commander-in-chief signals:

Heads of divisions E.

Divisional commanders signal: *Head of division E.;* and so soon as the column is formed to the Eastward, the commander-in-chief makes general signal *N.*, and the manœuvre is completed.

68. The fleet being in double column, heading N., to form it into line to the front.

1st Method.

Fig. 104.

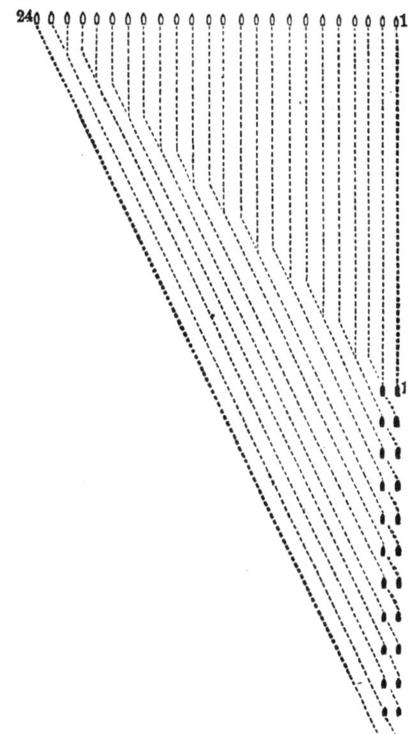

The commander-in-chief makes signal:

Forward into line—left oblique.

Flag-ship of van division: *Division N.N.W.—right vessels N.*

Flag-ships of centre and rear divisions signal: *Division N.N.W.*

The leading vessels continue onward under steerage-way; the other vessels keep two points to port at full speed. When three and four find one and two bearing from them on a line perpendicular to the course (E. is the bearing in this case), they so regulate their speed as to maintain this bearing until they are at the proper distance from them, when they come to the course under steerage-way, and hoist the position pennant as a signal to five and six, which now come into line in a similar manner.

The line may be formed according to the same principles, if the fleet be in triple column, column of fours, etc., etc.

2d Method.

Break the fleet into column of vessels as in manœuvre 37, and then into line as by manœuvre 66, 2d Method.

69. The fleet being in double columns abreast, by divisions, heading N., to form it into line to the front.

1st Method.

The commander-in-chief signals:

Forward into line—left oblique.

Flag-ships of divisions signal: *Division N.N.W.—right vessel N.*

Performed by each division as explained in manœuvre 68, 1st Method.

2d Method.

Break each division into column of vessels, and then proceed as in manœuvre 67, 2d Method.

70. The fleet being in column of vessels, heading N, to form it into line to the right or left, at right angles to the original direction.

Fig. 105.

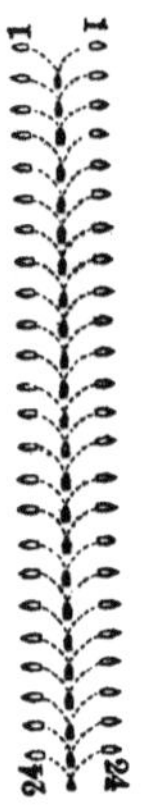

The commander-in-chief makes signal:

E. or *W.*

Which is repeated by the divisional commanders.

All come E. or W., and the fleet is in line.

Note.—We see by the diagram, that when the fleet, in natural order, turns to port, it comes into line in natural order; and when it turns to starboard it is in order reversed. Should the commander-in-chief desire to form it to starboard in natural order, or to port in order reversed, he has but to change the direction of the column to the opposite course before forming it into line. For instance, if he desire to form it to starboard in natural order, he first signals, *Head of fleet E.;* and so soon as he perceives that the leader of the fleet is turning to starboard, he signals, *Head of fleet S.* (See 56, Fig. 87.) When the fleet is in column to the Southward, he makes general signal *E.*, and the line is in natural order.

71. The fleet being in column of vessels abreast, by divisions, heading N., to form it into line to the right or left, at right angles to the original direction.

Break the fleet into column of vessels on the right or left division, as in manœuvre 42, Fig. 73, and afterward proceed as in manœuvre 70, Fig. 105.

72. The fleet being in double column, heading N., to form it into line to the right or left, at right angles to the original dirction.

FIG. 106.

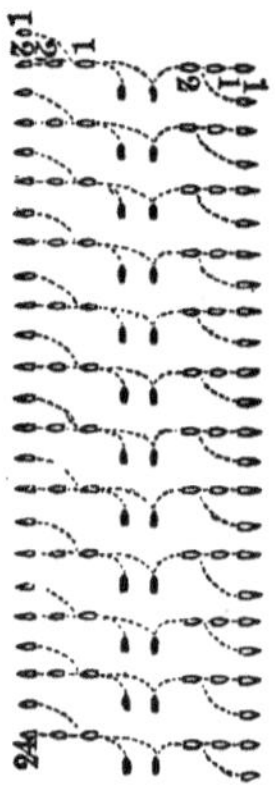

The commander-in-chief signals:

E. (or W.)

Which is repeated by the divisional commanders.

The fleet comes E. or W. at once, and moves by the flank in this direction, until the commander-in-chief makes signal, *Forward into line—right oblique;* when the vessels of the front line slow to steerage-way, while those of the rear line keep two points to starboard, and come up abreast of them, as shown by the diagram precisely as in forming double column. (See manœuvre 27, Fig. 56.)

It is evident that if the fleet were in triple column, column of fours, etc., etc., it could be formed into line to the right or left, at right angles to the original direction, according to the same principles.

73. The fleet being in double columns, triple columns, etc., etc., abreast, by divisions, to form it into line to the right or left, at right angles to the original direction.

Break each division into column of vessels, as in manœuvre 39, Fig. 70, then form the whole fleet into column of vessels on the right or left division, as in 42, Fig. 73, and finally proceed as in 70, Fig. 105.

To form the fleet from column into line *directly to the rear*, reverse it as in manœuvres 46, 47, 48, and 49, and then proceed as in *forming line to the front.*

74. The fleet being in column of vessels, to form it into line on any course whatever.

Form the column at right angles to the course on which you intend the *line shall head,* or, in other words, form it *on its line of bearing,* and then throw the column into line by making the compass signal. For instance, let the fleet be heading N., and suppose the line has to be formed to the S.E.

Fig. 107.

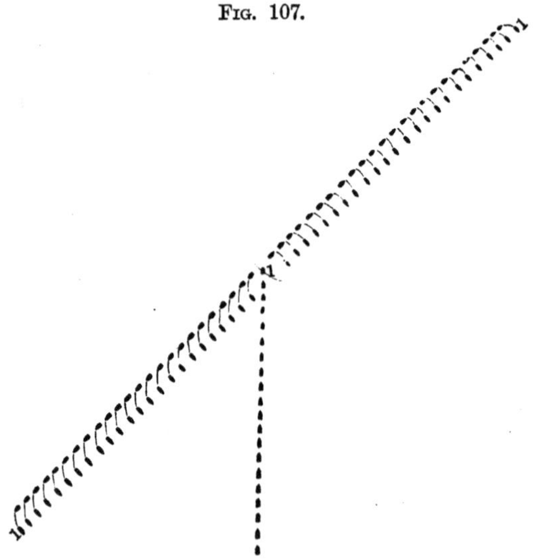

The commander-in-chief makes signal:

Head of fleet N.E. (or S.W.), and so soon as the column is formed on the course signalled, he signals S.E., and the fleet is in line heading S.E. on the line of bearing N.E. and S.W., as shown by the diagram.

If the fleet be in double column, or triple column, etc., etc., it must be broken into column of vessels and then manœuvred as above.

75. The fleet being in column of vessels abreast, by divisions, to form it into line on any course whatever.

Break the fleet into column of vessels on the right or left division, at right angles to the course on which the line is to head, as in manœuvres 43 and 46, Figs. 75 and 76, and then proceed as in manœuvre 74, Fig. 107.

If the divisions be in double column, or triple column, etc., etc., they must be broken into columns of vessels and then manœuvred as above.

76. The fleet being in double column, from the centre, heading N., to form it into line to the front.

FIG. 108.

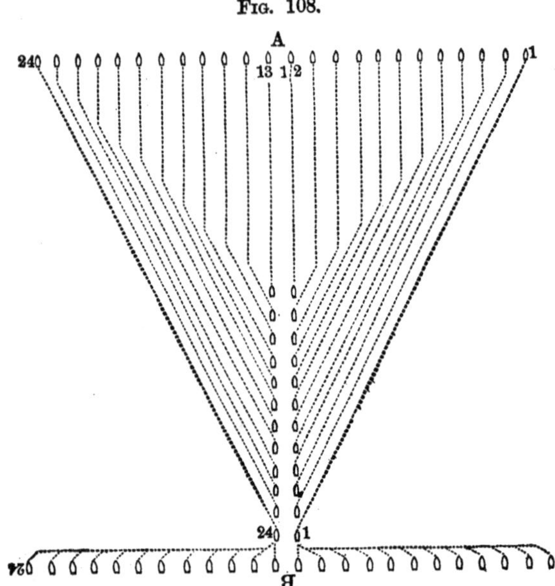

The commander-in-chief makes signal:

Forward into line.

Flag-ship of van division: *Division N.N.E.*

Flag-ship of rear division: *Division N.N.W.*

Flag-ship of centre division: *Van squadron** *N.N. E.—rear squadron N.N. W.—leading vessels N.*

Twelve and thirteen slow to steerage-way; the rest of the fleet oblique two points to starboard and port, as shown by the diagram, and come into line as explained in manœuvre 66, Fig. 100.†

To form the double column into line *to the rear*, the commander-in-chief reverses it (one and twenty-four becoming the leading vessels), and signals:

Forward into line, or head of starboard column W. —head of port column E.

Flag-ship of van division signals: *Head of division E.*

Flag-ship of rear division: *Head of division W.*

Flag-ship of centre division: *Division S.*

---

* The van squadron of the centre division is, of course, the leading squadron of that division when the fleet is in column of vessels in natural order—that is (in this case), numbers nine, ten, eleven, and twelve; the rear squadron being thirteen, fourteen, fifteen, and sixteen. As the distinguishing pennants of the vessels, squadrons, or divisions signalled are, however, always to be hoisted above the signal, no mistake can possibly occur.

† It is evident that this is analogous to the case of a column of vessels thrown into line to the front, the van and starboard half of the centre division (*the right wing of the fleet*) forming the starboard column of vessels, and the rear half of the centre division (or left wing of the fleet) forming the port column.

One and twenty-four alter course to E. and W. respectively, and are followed by the vessels of their respective columns. So soon as the commander-in-chief observes that twelve and thirteen are about coming into column, he signals *S.*, and the manœuvre is completed as shown by the diagram B.

An inspection of Fig. 32 will show that a fleet in column of divisions from the centre would be formed into line to the front or rear in a similar manner.

77. The fleet being in double columns from the centre of divisions, heading N., to form it into line to the front.

Fig. 109.

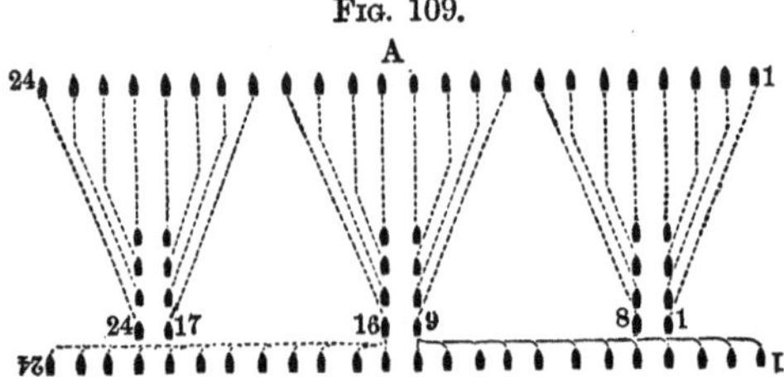

The commander-in-chief signals:

Forward into line.

Divisional commanders signal: *Van squadron N.N.E.—rear squadron N.N.W.—leading vessels N.*

Performed by each division as explained in 76, and shown by the diagram A.

To form the fleet into line to the rear, the commander-in-chief would reverse it and signal:

*Forward into line.**

While the divisional commanders would signal to the leading vessels of their divisions to keep E. and W., and come into line as explained in 76, and shown by the diagram B.

---

* Or, *Head of port columns E.—head of starboard columns W*

78 The fleet being in double column on the centre, heading N. to form it into line to the right or left, at right angles to the original direction.

1*st. Method.*

FIG. 110.

The commander-in-chief makes signal:

Fleet—form column of vessels—van in front.

Flag-ships of van and rear divisions signal: *Division N.*

Flag-ship of centre division signals: *Head of van* squadron N.N.W.—rear squadron N.*

The port column slows to steerage-way; the starboard column continues at full speed, its leader keeping *N.N.W.* at once and coming into column ahead of thirteen. So soon as eleven comes abreast of twelve she keeps N.N.W. and comes into column ahead of twelve ; ten manœuvres in a similar manner with respect to eleven, and so on to the last vessel.

When the fleet is in column of vessels, the commander-in-chief makes signal:

E. or W.

The line is formed as in 70, Fig. 105.

2d Method.

Form the fleet into line to the front, as in 76, Fig. 108 ; then into column, as in 1, Fig. 18, and finally signal E. or W., and the manœuvre is completed.

---

* See note (*) to manœuvre 76.

3d Method.

FIG. 111.

The commander-in-chief makes signal:

Starboard column E.—head of port column E.

Flag-ship of van division signals: *Division E.*

Flag-ship of centre division signals: *Van (or right centre) squadron E.—head of rear (or left centre) squadron E.**

Flag-ship of rear division: *Division N.*

So soon as the commander-in-chief observes that the starboard column is turning to the Eastward, he signals to it *S.*, and also, *Head of port*

* See note (*) to manœuvre 76.

column S., and the column is formed as shown by the diagram, from which it can of course be formed into line to the right or left by the signal *W. or E.*

To form a fleet, in column of divisions from the centre, into line to the right or left, at right angles to the original direction, form it first into line to the front, then into column of vessels from the right or left, as in manœuvre 1, Fig. 18, and finally proceed as above.

79. The fleet being in double columns from the centre of divisions, heading N., to form it into line to the right or left, at right angles to the original direction.

Form the fleet into line, as in manœuvre 77, Fig. 109; then into column from the right or left, as in manœuvre 1, Fig. 18, and finally into line to the right or left, upon the signal *E. or W.*

80. The fleet being in double columns from the centre, to form it into line on any course whatever.

Form it into column of vessels by one of the methods described in manœuvre 78; then into column of vessels on the *line of bearing of the line,* as in manœuvre 74, Fig. 107, and finally into line by signalling the course, as in manœuvre 74, Fig. 107.

81. The fleet being in double columns from the centre of divisions, to form it into line, on any course whatever.

Form it into line to the front, as in manœuvre 77, Fig. 109 ; then into column of vessels from the right or left, at right angles to the course on which the line is to head (or, in other words, on its line of bearing), and finally into line by signalling the course, as in manœuvre 74, Fig. 107.

82. The fleet being in line, heading N., to form it into echelon of vessels from the right, preserving the original direction.

1*st Method.*

FIG. 112.

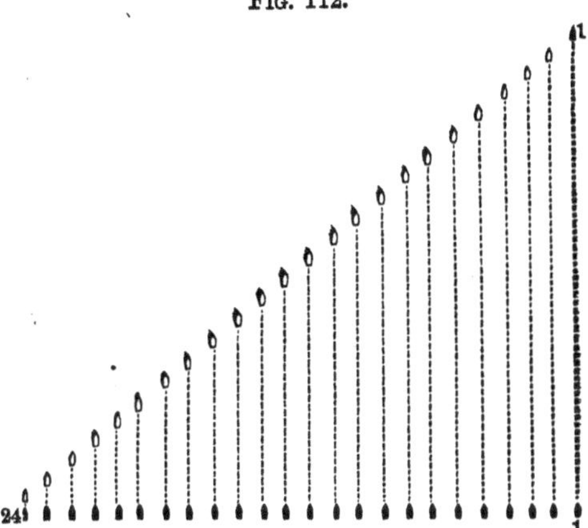

The commander-in-chief makes signal:

From the right of fleet, form echelon of vessels.

Which is repeated by the divisional command-

The vessel on the right of the fleet steams at full speed; the other vessels slow to steerage-way. When two finds one bearing N.E. from her, she steams at full speed ; three steams at full speed when two bears N. E. from her, and so on to the last vessel.

Echelon of vessels, from the left, may be formed according to the same principles, upon the signal: *From the left of fleet, form echelon of vessels.*

2d Method.

Form the fleet into column of vessels, from the right or left, to the N.E., S.W., N.W., or S.E.; then signal *N.*, and the manœuvre is completed.*

83. The fleet being in line, heading N., to form it into echelon of vessels, on the starboard or port bow or quarter.

FIG. 113.

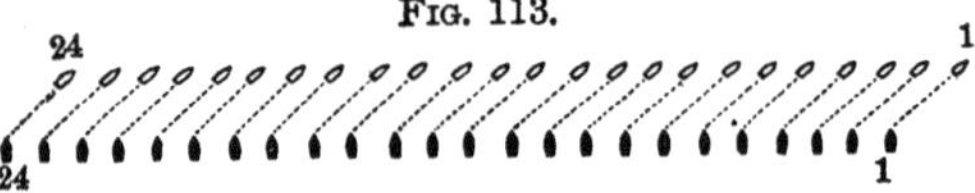

To form it on the starboard bow, the commander-in-chief makes signal :

N. E.†

* See diagram. In this case, of course, the line of bearing will be N.E. and S.W., but if formed from the left, it would be N.W. and S.E.

† It is very necessary here that the vessels shall describe equal arcs in turning.

Which is repeated by the divisional commanders.

The vessels alter course together to N.E., and are in echelon on a line of bearing *E.* and W.

To form the fleet into echelon on the starboard quarter, the commander-in-chief first forms it into columns of vessels to the right (as in 10, Fig. 35), and then into echelon by signalling a course to starboard, forming an angle of 45° with the column. In this case the course signalled would be S.E., the line of bearing remaining E. and W.

Inversely, the fleet may be formed into echelon on the port bow or quarter, according to the same principles.

84. The fleet being in line, to form it into echelon of vessels on any course whatever.

Form it into column of vessels, from the right or left, *on what is to be its line of bearing when in echelon;* then signal the course to be steered, and the manœuvre is completed. Suppose, for example, the fleet to be in line, steering N., and it be required to form it into echelon of vessels heading E.N.E.

FIG. 114.

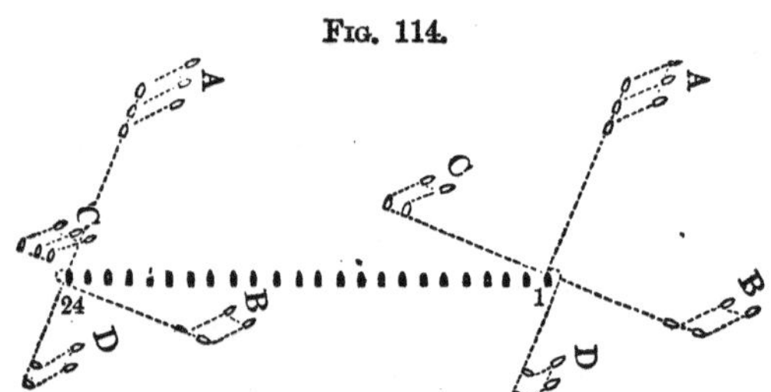

The commander-in-chief forms the fleet into column of vessels, from the right or left, to the N.N.E., and then signals *E.N.E.*, and the fleet is in echelon on the line of bearing N.N.E. and S.S.W., as shown by A.

Or he may form it into column of vessels, from the right or left to the E.S.E., and signal *E.N.E.*, and the fleet will be in echelon on the line of bearing E.S.E. and W.N.W. (See B.)

Or into column of vessels to the W.N.W., then into line to the N.N.E., and finally make signal *E.N.E.*, and the fleet will be in echelon on the line of bearing W.N.W. and E.S.E. (See C.)

Or into column of vessels to the S.S.W., then into line to the E.S.E., and finally make signal *E.N.E.*, and the fleet will be in echelon on the line of bearing S.S.W. and N.N.E. (See D.)

85 **The fleet being in line, to form it into echelons of vessels, from the right of divisions, preserving the original direction.**

1st Method.

FIG. 115.

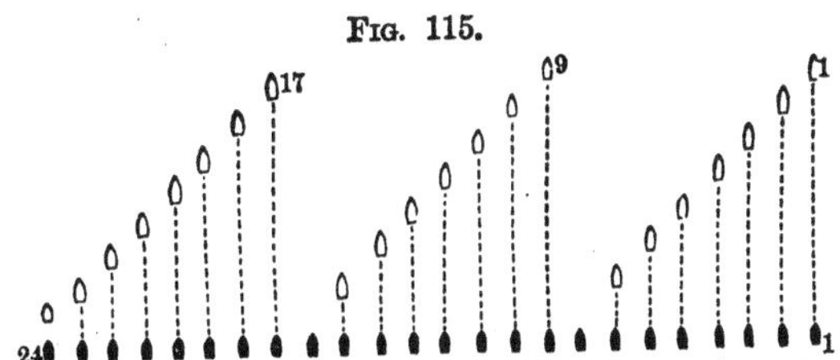

The commander-in-chief makes signal:

From the right of divisions, form echelons of vessels.

Divisional commanders signal: *Division—from the right, form echelon of vessels.*

The vessels on the right of divisions continue onward at full speed; the other vessels slow to steerage-way. When two, ten, and eighteen find one, nine, and seventeen bearing N.E. from them respectively, they go ahead at full speed, and so on to the last vessels.

Echelons of vessels, from the left of divisions, may be formed according to the same principles.

2d Method.

Form the divisions into columns of vessels, from the right or left, to the N.E. or N.W.; then signal N., and the manœuvre is completed.

86. The fleet being in line, to form it into echelons of vessels, from the right or left of divisions, on any course whatever.

Form the fleet into columns of vessels, from the right or left, at right angles to the course which the vessels are to steer when in echelon; next into line by *signalling this course*, and then form the divisions into echelons of vessels, as in 85

Suppose, for example, the fleet to be in line, steering N., and it be required to form its divisions into echelons of vessels heading S.E.

FIG. 116.

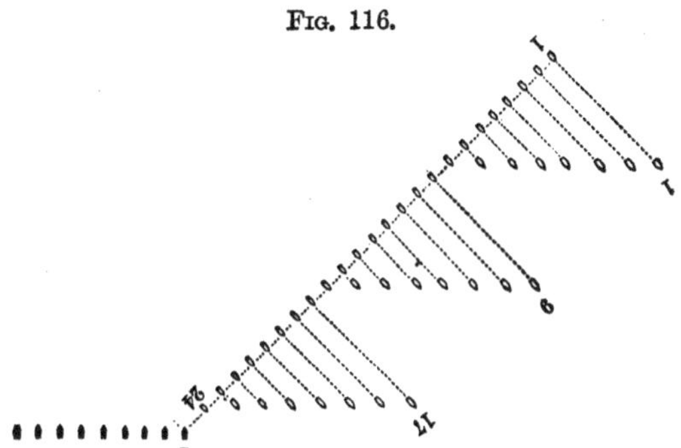

The fleet is first formed into column of vessels to the N.E.,* as in manœuvre 13, Fig. 40; next into line to the S.E., upon general signal *S.E.*, and finally into echelons of vessels from the left† of divisions, as shown by the diagram.

---

* Or it may be formed to the *S. W.*

† That it may be formed into echelons of vessels from the right of divisions, as well, is plain.

87 The fleet being in line, heading N., to form it into echelon of squadrons, from the right, preserving the original direction.

Fig. 117.

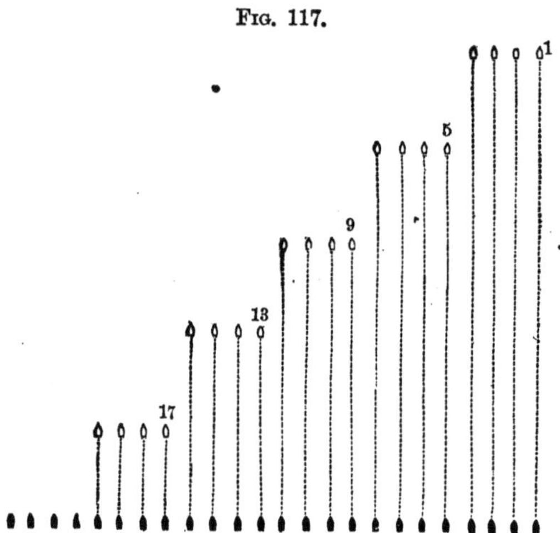

The commander-in-chief makes signal:

From the right of fleet, form echelon of squadrons.

Which is repeated by the divisional commanders.

The squadron on the right of the fleet steams at full speed; the other squadrons slow to steerage-way. When the second squadron finds the first bearing from it at an angle of 45° from the course N.E. is the bearing in this case), it proceeds at full speed; and so on to the last squadron.

It is evident that the fleet can be formed into echelon of squadrons from the left, and into echelon of divisions from the right and left, according to the same principles.

88. The fleet being in line, heading N., to form it into echelons of squadrons, from the right of divisions, preserving the original direction.

FIG. 118.

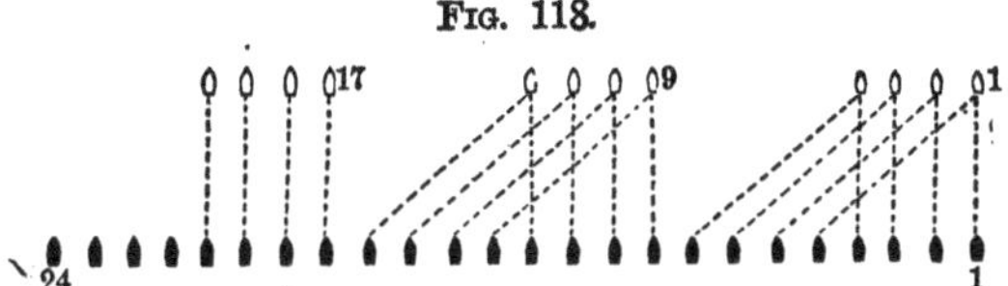

The commander-in-chief makes signal :

From the right of divisions, form echelons of squadrons.

Divisional commanders make signal : *Division —from the right, form echelon of squadrons.*

The squadron on the right of each division keeps at full speed ; the other squadrons slow to steerage-way, resuming their speed when the right squadrons bear from them at an angle of 45° (N.E. in this case) from the course.

It is evident that the fleet can be formed into echelons of squadrons, from the left divisions, according to the same principles.

89. The fleet being in line, to form it into echelons of squadrons or divisions, from the right or left, on any course whatever.

Form the fleet into column of vessels, from the right or left, at right angles to the course it is to steer when in echelon ; next into line *by signalling this course,* and then proceed as in 87.

90. The fleet being in line, to form it into echelons of squadrons, from the right of divisions, on any course whatever.

Form the fleet into column of vessels, from the right or left, at right angles to the course it is to steer when in echelon; next into line *by signalling this course,* and then proceed as in 88.

91. The fleet being in line, heading N., to form it into double echelon from the centre, preserving the original direction.

Suppose, for example, it be required to form it into double echelon from thirteen.

FIG. 119.

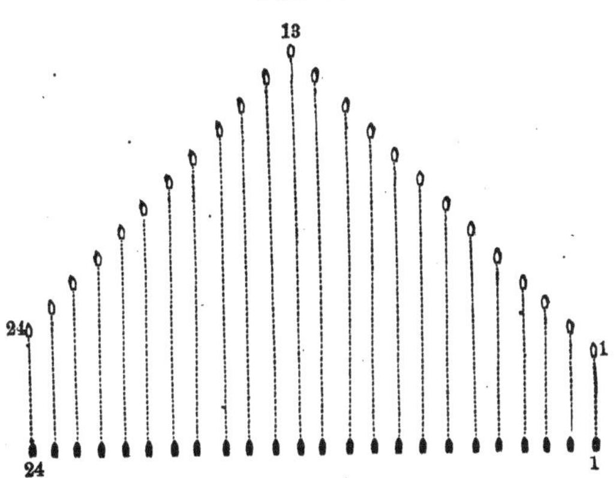

The commander-in-chief makes signal:

From the vessel whose distinguishing pennant is shown above this signal, form double echelon.

Which is repeated by the divisional commanders.

The left centre vessel (thirteen) steams at full speed; the other vessels slow to steerage-way.

When twelve and fourteen find thirteen bearing from them, respectively, N.W. and N.E. (that is, at an angle of 45° with the course N.), they steam at full speed; eleven and fifteen steam at full speed, when twelve and fourteen bear from them, respectively, N.W. and N.E.; and so with the remaining vessels.

It is evident that the fleet may be formed into double echelon from the right centre vessel, or from the two centre vessels, according to the same principles.*

92. The fleet being in line, heading N., to form it into double echelons from the centre of divisions, preserving the original direction.

Suppose, for example, it be required to form each division into double echelon on its left centre vessel.

FIG. 120.

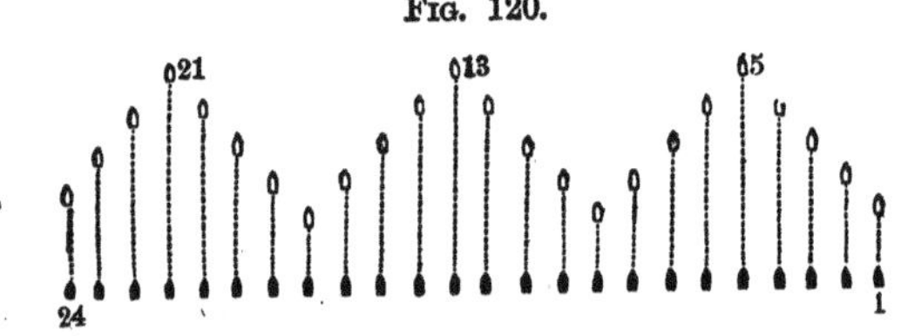

* Of course thirteen's pennant is hoisted above this signal. See Index of Signals.

The commander-in-chief makes signal:

*From the left centre of divisions, form double echelon.**

Divisional commanders signal: *Division—from the left centre vessel, form double echelon.*

In each division the left centre vessel continues at full speed; the vessels on the right and left of her slow to steerage-way until she bears from them, respectively, N.W. and N.E., when they resume their speed.

It is evident that the fleet may be formed into double echelon from the right centre vessel, or from the two centre vessels of divisions, according to the same principles.

---

* In making this signal, the commander-in-chief runs up the distinguishing pennants of five, thirteen, and twenty-one over number 69 (see Index of Signals), while each divisional commander hoists number 69 under the distinguishing pennant of the vessel of his division, which has been designated by the commander-in-chief.

93. The fleet being in line, heading N., to form it into double echelon from the flanks,* that is from the van and rear vessels (or vessels on the extreme right and left), preserving the original direction.

Suppose, in this case, the commander-in-chief desires to form it from the flanks, *on the left centre vessel.*†

FIG. 121.

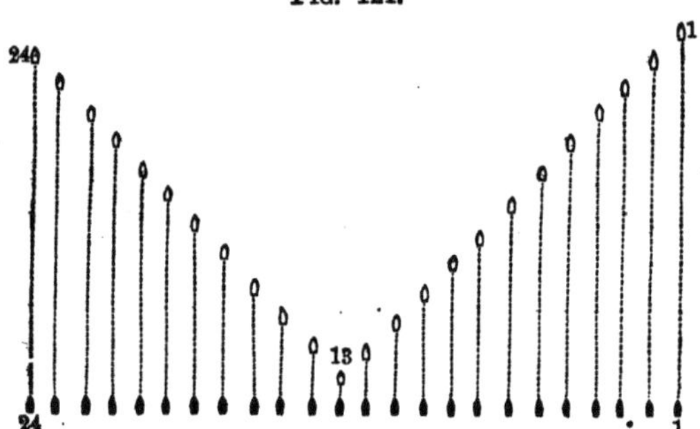

The commander-in-chief makes signal:

From the right and left of fleet—on the left centre vessel‡*—form double echelon.*

---

* With the wings in advance, in other words.

† If ordered to be formed on the right centre, the vessel on the *extreme left* would be the first to move forward; if on the two centre vessels, the vessels on the extreme right and left of the fleet would continue at full speed, as is evident.

‡ The commander-in-chief here hoists thirteen's distinguishing pennant over number 70. (See Index of Signals.)

Which is repeated by the divisional commanders.

One continues onward at full speed; the other vessels slow to steerage-way. When two finds one bearing from her at an angle of 45° from the course (N.E. in this case), she hoists the position pennant as a signal to the vessel on the extreme left, and with this vessel resumes full speed; three and twenty-three steam at full speed when two and twenty-four bear N.E. and N.W. from them, respectively. The other vessels manœuvre in a similar manner; thirteen resuming full speed, when twelve and fourteen bear from her N.E. and N.W., respectively.

Double echelon may be formed from the flanks on the right centre vessel, or on the two centre vessels of the fleet, according to the same principles.

94. The fleet being in line, heading N., to form it into double echelons from the flanks of divisions, on their left centre vessels, preserving the original direction.

Fig. 122.

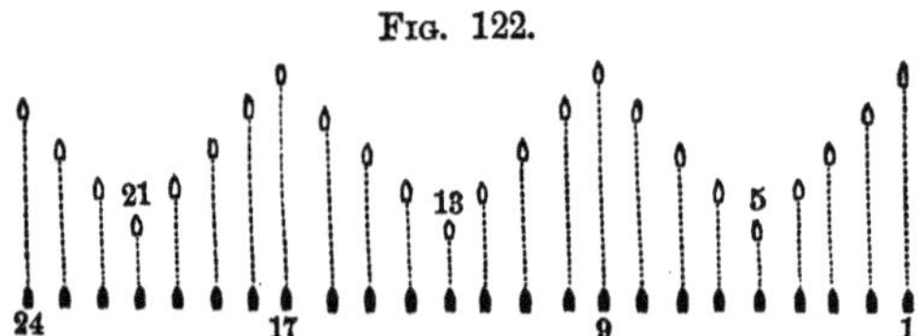

The commander-in-chief makes signal:

*From the right and left of divisions—on the left centre vessels—form double echelons.**

The divisional commanders signal: *Division—from the right and left—on the left centre vessel form double echelon.**

The vessels on the right of divisions (one, nine, and seventeen) continue at full speed; the other vessels slow to steerage-way. When two, ten, and eighteen find one, nine, and seventeen bearing from them, respectively, at an angle of 45° with the course (N.E. in this case), they hoist

* The commander-in-chief here hoists the distinguishing pennants of five, thirteen, and twenty-one over number 70 (see Index of Signals), while each divisional commander hoists the same number under the distinguishing pennant of the vessel of his division designated by the commander-in-chief.

the position pennant as a signal to the vessels on the extreme left of divisions (eight, sixteen, and twenty-four), and with these vessels resume full speed. When three and seven, eleven and fifteen, and nineteen and twenty-three find two and eight, ten and sixteen, and eighteen and twenty-four bearing from them, respectively, N.E. and N.W., they resume their speed. The other vessels manœuvre in a similar manner, five, thirteen, and twenty-one going at full speed, when the vessels next on their right and left bear, respectively, from them N.E. and N.W.

Double echelon may be formed from the flanks, on the right centre vessel or on the two centre vessels of divisions, according to the same principles.*

* See note (*) to manœuvre 93.

95. The fleet being in line, heading N., to form it into double echelon of squadrons from the left centre squadron, preserving the original direction.

FIG. 123.

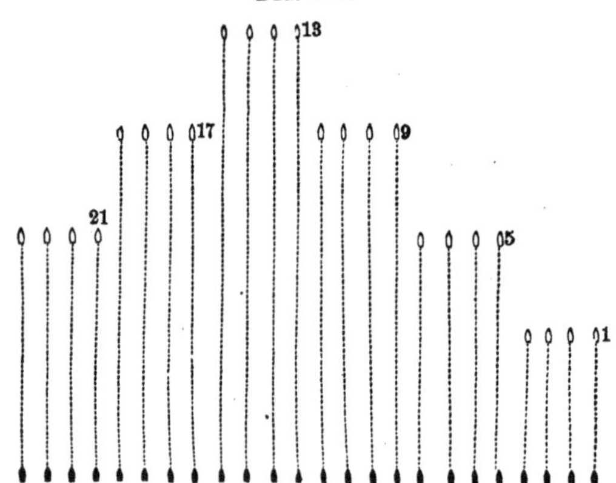

The commander-in-chief makes signal:

From the left centre squadron, form double echelon of squadrons N.*

Which is repeated by the divisional commanders.

The left centre squadron keeps its speed, the other squadrons slow to steerage-way. When the third and fifth squadrons find the left centre

* Distinguishing pennant of this squadron is here hoisted over number 71. See Index of Signals.

squadron bearing from them, respectively, N.W. and N.E., they resume their speed, etc., etc.

Inversely, the fleet may be formed into double echelon of squadrons from the flanks, on one of the centre squadrons, upon the signal :

From the right and left squadrons—on the—— squadron—form double echelon of squadrons.*

96. The fleet being in echelon of vessels, heading N., to form it into double echelon, on the left centre vessel with the wings in advance, and preserving the original direction.

FIG. 124.

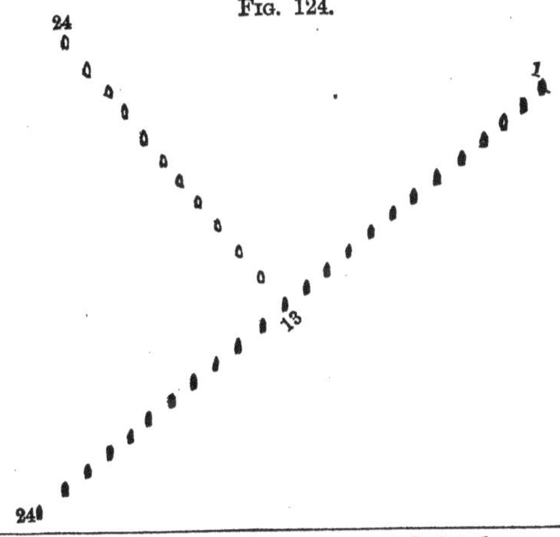

* Squadron's distinguishing pennant hoisted over number 72. See Index of Signals.

The commander-in-chief makes signal:

On the left centre vessel, form double echelon—left wing forward.*

Which is repeated by the divisional commanders.

All the vessels of the right wing (thirteen included) slow to steerage way; the other vessels maintain their speed, fourteen slowing when bearing N.W. from thirteen, fifteen when bearing N.W. from fourteen, etc., etc. When the vessel on the extreme left hoists the *position pennant,* the fleet resumes its speed.

It is evident that it can be formed into double echelon on the right centre vessel, or on the two centre vessels, according to the same principles.

Inversely, the fleet may be formed from double echelon into echelon of vessels in the same manner.

---

* Thirteen's distinguishing pennant is here hoisted above number 74. See Index of Signals.

97. The fleet being in echelon of vessels, heading N., to form it into double echelon from the left centre vessel (that is with this vessel leading), preserving the original direction.

FIG. 125.

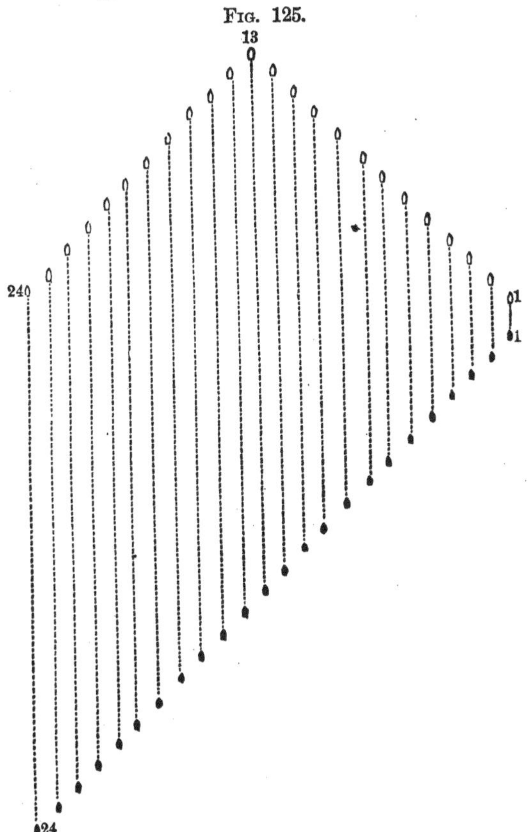

The commander-in-chief makes signal:

From the left centre vessel, form double echelon—right wing to the rear*

Which is repeated by the divisional commanders.

The left wing continues its speed; the right wing slows to steerage-way, twelve resuming her speed when thirteen bears N.W. from her, eleven when twelve bears N.W. from her, etc., etc., etc.

Inversely, according to the same principles a fleet in double echelon from the centre may be formed into echelon of vessels.

98. The fleet being in echelon of vessels, to form it into column of vessels.

The commander-in-chief has but to signal the line of bearing and the column will be formed.

For example, take the fleet heading N. with the right wing in advance (as in Fig. 124), and we see that if the commander-in-chief should make general signal N.E. the fleet would be in column of vessels with the van (or right) in front; should the signal be S.W. the column would be the same, with the rear (or left) in front.

---

* Thirteen's distinguishing pennant must be here hoisted over number 75. See Index of Signals.

To form a fleet from double echelon into column of vessels, break it first into echelon of vessels and then proceed as above.

2d Method.

Let the fleet be in double echelon on the left centre vessel (as in Fig. 121) and the commander-in-chief desire to break it into column of vessels with the van (or right) leading, and we see that he has but to make general signal N.E., and so soon as the whole fleet has obeyed the signal, to follow it up with: *From the right, form column of vessels,* and the manœuvre will be completed. For the right wing will have formed column so soon as it shall have obeyed the signal to steer N.E., and the left wing will come into column as in manœuvre 1, Figs. 16, 17, and 18.

If, in this case, the commander-in-chief should desire to form the fleet into column with the rear (or left) leading, he would first make general signal *N. W.*, and afterward, *From the left, form column of vessels.*

99. The fleet being in echelon of vessels, heading N., to form it into line to the front.

Fig. 126.

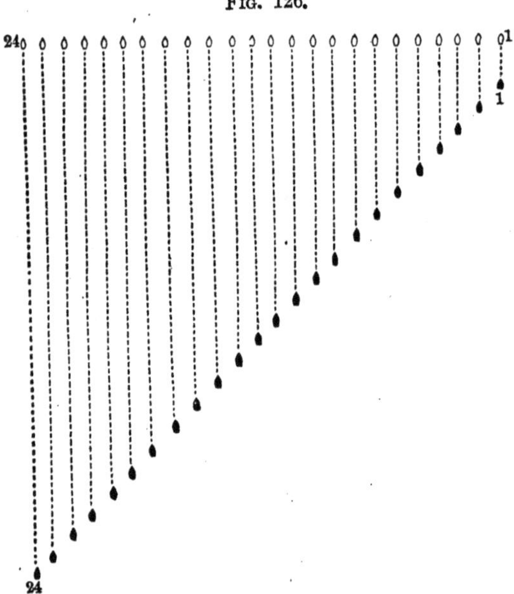

The commander-in-chief makes signal:

Forward into line.

Which is repeated by the divisional commanders.

One slows to steerage-way; the other vessels maintain their speed, slowing in succession, as they get abeam of their leader. When the line is formed, the fleet moves at a uniform rate of speed.

To form the line to the E. or W. (that is, at right angles to the original direction), the commander-in-chief first signals *E.* or *W.*, and afterward, *Forward into line.*

To form the line to the rear, he would signal *S.*, and afterward, *Forward into line.*

A fleet in echelon of vessels, by divisions or squadrons, or in echelon of squadrons, would be thrown into line to the front or rear, according to the same principles.

100. The fleet being in echelon from the centre, heading N., to form it into line to the front.

Fig. 127.

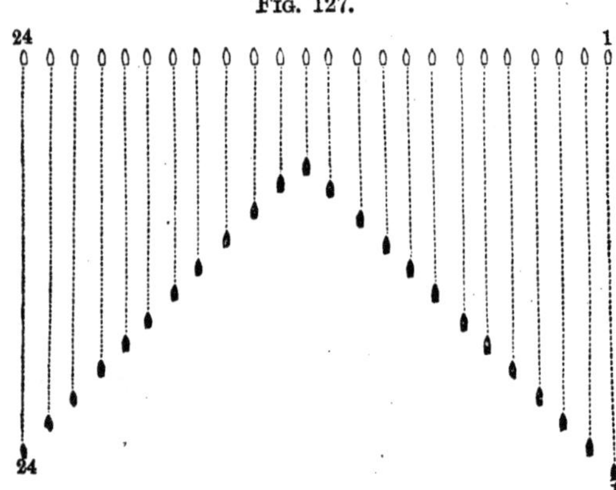

The commander-in-chief makes signal:

Forward into line.

Divisional commanders signal: *Division—forward into line.*

The leading vessel slows to steerage-way; the other vessels keep their speed, each one slowing as she gets up abeam of the leader. When the line is formed, the fleet moves at a uniform rate of speed.

If the fleet were in double echelon from the

flanks on the centre, the vessels on the extreme right and left would be the first to slow to steerage-way, and the line would be formed on them, as is evident.

To form the line to the right or left, the fleet should be broken into echelon of vessels and afterward thrown into line, as in manœuvre 99.

To form it to the rear in this case, the commander-in-chief would first signal *S.*, and afterward, when the fleet was heading S., *Forward into line.*

101. The fleet being in line, to form it in two lines in the order of battle.

Fig. 128.

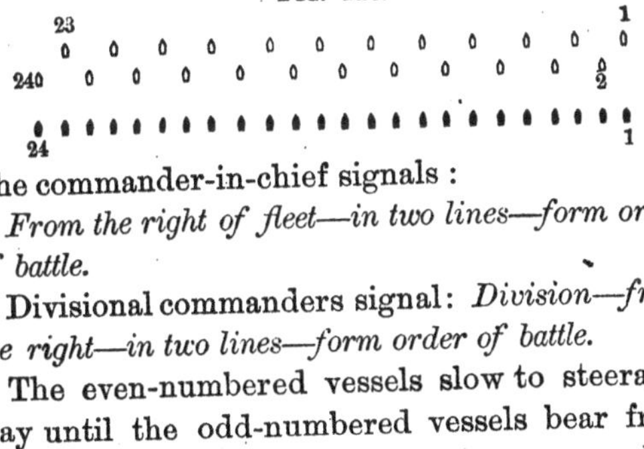

The commander-in-chief signals :

From the right of fleet—in two lines—form order of battle.

Divisional commanders signal: *Division—from the right—in two lines—form order of battle.*

The even-numbered vessels slow to steerage-way until the odd-numbered vessels bear from them at an angle of 45° with the course, when they resume their speed. For instance, two resumes full speed when one bears N.E. from her (supposing the fleet to be steering N.), four when she brings three on this bearing, etc., etc., etc.

If the signal were: *From the left of fleet—in two lines—form order of battle,* the even-numbered vessels would maintain their speed and the others slow to steerage-way.

Should the commander-in-chief desire to form from two lines into one again, he has simply to signal, *Forward into line,* when the vessels of the rear line will move up and fill the gaps in the front line.

102. The fleet being in double column, to form it into order of battle in two columns.

FIG. 129.

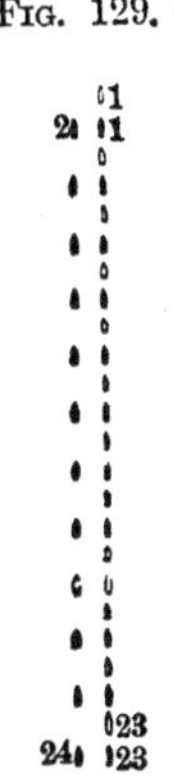

The commander-in-chief makes signal:

From the right—in two columns—form order of battle.

Divisional commanders signal: *Division—from the right—in two columns—form order of battle.*

The port column slows to steerage-way, each vessel of it resuming her speed when her consort has changed her bearing from abeam to an angle of 45° with the course.

If the signal were: *From the left—in two columns*

—*form order of battle,* the port column would maintain its speed, and the starboard one slow to steerage-way.

Should the commander-in-chief desire to re-form into double column, he has simply to make signal, *Form double column,* when the vessels of the advance column will slow until their consorts have got abreast of them.

An inspection of diagrams 128 and 129 will show that by signalling a course at right angles to the one steered by the fleet, the commander-in-chief can form it from the order of battle in two lines to the order of battle in two columns, or from the latter disposition to the former.

Upon assembling a fleet the commanding officer should select the three longest vessels as *regulators* for its three divisions, and detail a competent officer to experiment with them, thus: First, on a calm day, form them into column of vessels in "close order," heading N., and signal a speed of eight knots an hour. So soon as this speed has become uniform, watch a time when they are on *the exact line of bearing N. and S.*, and signal *E.* All now put their helms hard a-port and find themselves abreast, heading E, *without having changed their line of bearing, provided all have described equal arcs in turning.* If they have not, the experiment must be repeated until they do, the vessel which at first described the longest arc, now alone putting her helm hard over, and the other vessels so graduating their helms as to make their arcs of manœuvre correspond with hers. In the same manner, each division is exercised with its "regulator," and much valuable information is gained thereby by each commanding officer, as to the *absolute* and *relative turning power* of his vessel. This, however, is simply to be regarded as an exercise for the instruction of the officers, as before stated, and by no means to be established as a rule of manœuvre: for the circles described by two vessels whose helms are so graduated as to enable them to describe precisely similar arcs, in calm weather and when not affected by currents, will vary with every changing breeze and tide, and no absolute rule, in this regard, can be established *practically;* nor is it desirable that it should be. To lay it down *theoretically* is, in reality, to restrict the *turning power* of a fleet to that of its *longest* and *clumsiest* vessel.

In the Potomac flotilla, whose vessels ranged from eighty to three hundred feet in length, my custom was, in manœuvring, to direct that each vessel should turn *as quickly as possible,* experience soon demonstrating that the *long* vessel, when *steadied on her course,* quickly recovered, through superior swiftness, the distance *she had lost in turning.* It is to be hoped, however, that by the aid of twin screws or the application of hydraulic pressure, the time is not far distant when vessels shall be made to turn almost about their centres.

When a vessel has been signalled to act as a guide for the fleet, it is the duty of her commanding officer to give his whole attention to her steerage, since great confusion must inevitably result from any negligence in this regard. So long as the guide vessel is on her course, her guide flag should be kept well rounded up to the mast-head; but should she from any cause deviate from it, the guide flag must be lowered a few feet, care being taken to rehoist it so soon as the vessel has resumed her course.

The evolutions of a sailing fleet must necessarily be performed by successive movements on the part of its vessels; but to prescribe successive movements for a fleet of *steamers* is absurd. For example, the vessels of a steam fleet in line heading N., and ordered to form column of vessels from the right, heading N.E., would, upon the hauling down of the flags of manœuvre, put their helms *simultaneously* to port, the leader of the column righting her helm when nearing her course N.E., and the others swinging to starboard *together* until heading E. This is the only mode of insuring precision in the performance of evolutions, and in a series of manœuvres conducted on this principle and extending over two years (often in blowy weather), with from six to sixteen vessels, no collision occurred.

As it is of the utmost importance to each vessel to know the speed of the vessels nearest to her, all the fleet should keep "bent on" a red ball,* to be used as a speed signal, thus:

* At night, in time of peace, a lantern may be substituted.

Ten knots and over........	Ball out of sight on deck.
From eight to ten knots....	" resting on the rail.
" six to eight knots	" a few feet above the rail.
" four to six knots	" at half mast.
Engine just turning over ...	" a few feet below the truck.
" " stopped........	" mast-headed.

Any injury to the motive or steering power of a vessel should be at once signalled to the divisional commander, and by him communicated to the commander-in-chief.

At night, a large, red lantern should be suspended over the stern, having over it a tarpaulin cover. To this cover a lanyard should be attached leading on deck, and a man stationed by it. The instant the engine stops the cover is to be removed, and, at the same time, two short, sharp blasts with the steam whistle sounded to call attention to the signal, which must be answered by the vessel astern with one blast. Should the lantern, when uncovered, be lowered to the water's edge, it will signify that the vessel has stopped because a man is overboard, when it will be the duty of the vessel astern to lower her boats and assist in the search for him. When the lantern is hoisted up again, it will be a signal that the man has been picked up, or search for him abandoned; and, when re-covered and one long blast sounded, that the vessel has resumed her headway.

Each vessel should maintain her proper distance and bearing with reference to the "guide vessel," rather than her next ahead, in "column;" and, in "line," with the "guide vessel," rather than her next to starboard (if the guide be right) or her next to port (if the guide be left). Consequently, when a vessel "falls out" of line or column, her place is to be left vacant until the squadron or divisional commander shall signal to the contrary; and the vessel losing her place is to regain it as speedily as possible, unless another station be assigned to her. *During action, however, in the absence of signals to the contrary, every gap is to be filled as speedily as possible.*

When a guide vessel is disabled, her next astern, or to port or starboard (according as the guide is right or left), shall immediately hoist the "guide flag," unless signal be made to the contrary.

From sunset to sunrise, under all circumstances, and during the day in heavy or thick weather, to avoid the danger of collisions, the fleet should steam in open order; and an echelon formation is recommended.

The greatest speed signalled by the admiral

to the fleet should be a half-knot less than that which its slowest vessel is capable of maintaining, in order that this vessel may be enabled to "close up," should she, from any cause, lose distance.

Vessels engaged in carrying despatches or orders should hoist, *below* their despatch flag, the distinguishing pennants of the vessels to which they are bound, *in the order in which they intend to communicate with them, counting from the despatch flag downward.* In this way the transmission of intelligence may be greatly facilitated (for the vessels to be last boarded, knowing whither the despatch vessel is bound, can readily send their boats to intercept her); and so soon as a despatch has been delivered for a vessel, her distinguishing pennant must be hauled down.

If a vessel be disabled, it becomes the duty of her next astern to tow her, provided the fleet be in column of vessels; but if the fleet be steaming in any other order, this duty belongs to her next to port, unless the disabled vessel be on the left flank, when it devolves upon her next to starboard.

A vessel coming suddenly upon danger should signal it, and the course necessary for the fleet to steer, in order to avoid it. Thus, supposing a reef to be discovered, and it be necessary for the fleet to steer N.W. to clear it, the vessel discovering it signals: *Danger !—fleet N. W.*, when the fleet comes N.W. instantly, and keeps this course until another is signalled by the admiral.

When a fleet anchors in a narrow or crowded harbor, van (or right) in front, it should be got under way rear (or left) in front, and *vice versa.*

Although the fleet is not to be governed by the course steered by the admiral, yet, in all else, he is to be closely followed by his divisional officers, whose motions, in turn, must be followed by the vessels of their respective commands. Thus, if the admiral go to quarters, all the fleet go to quarters; if he spread awnings, all the fleet spread awnings, etc., etc., etc.

In the days when ships were at the mercy of the winds, an admiral could not do better than *lead* his fleet into "close action;" but now that, through the agency of steam, war has become not less a science at sea than on land; when the ocean is a great chess-board, upon which the skilful looker-on sees many a move not apparent to the contestants, whose brains have become heated with the strife, the *role* of the admiral approximates to that of the general, and he should, like the latter, take post whence, without being an active participant in it, he may overlook the whole *sea of battle,* and signal to the fleet such formations as he shall find necessary. He should, in other words, be the *mind* of the fleet, and his officers and men should feel that he is watching over them, ever ready to take advantage of a false move on the part of the enemy.

"Soldiers," said Napoleon at Austerlitz, "I will myself direct your battalions. I will keep myself at a distance so long as, with your accustomed valor, you carry disorder and confusion into the enemy's ranks; but, should victory appear for a moment uncertain, you shall see your Emperor expose himself." This is precisely the relation in which an American admiral should stand toward his seamen. When, after having formed such tactical combinations as he has deemed judicious for carrying out his plan of attack or order of battle—after having disposed of the reserve and put every gun into action—he shall feel that victory is doubtful, then, and *not till then*, let him bear down into the thickest of the fight, and "encourage, *in his own person*, his officers and men to fight courageously."

The commander of a vessel of war is intrusted with the honor of the flag which flies at her mast-head, and with the lives of her officers and crew. His duty to the former requires him to engage every enemy he may fall in with of equal force, and to continue the engagement while he has power to do so. His duty to the latter and the dictates of humanity equally require that, when *resistance is no longer possible,* he shall haul down his colors and confess himself vanquished. This is the law of single combats.

In fleet fighting, however, the case is widely different. Each vessel is now but a part of a whole, and cannot be considered whipped until the whole fleet is beaten. It becomes, therefore, the imperative duty of each captain in a fleet to fight his vessel while he has a gun to fire, a bow to ram with, or a deck to stand upon; for, however crippled he may be, so long as he fights, he is rendering good service to his comrades, *by keeping an enemy from turning his guns on them.*

If disabled in machinery, he should (when to windward) make all sail and bear down on an enemy for the purpose of disabling him, or carrying him by boarding, always keeping in mind that he might as well be captured *as to go out of action without taking an opponent with him.*

Every captain in battle knows his own wounds, but is ignorant of his enemy's, which may be much more serious; and many cases might be cited where well-timed obstinacy in surrendering would have made of the vanquished the victor. "Victory," said Napoleon, "belongs to the most persevering."

In the presence of an enemy *under way*, the fleet should be kept under full steam, as, otherwise, it would be in the enemy's power to concentrate his whole force upon one portion of it, and thus destroy it in detail.

In a *fair stand-up* fight with forts or batteries, only such steam should be carried (to avoid the disaster incident to a boiler being struck under a high pressure of steam) as may be deemed necessary by each captain for the easy handling of his vessel.

When the object is simply to run by forts or batteries, however, in an unobstructed channel, go at full speed, as it is better to take the *remote chance* of a boiler explosion from shot, than to incur the *certain danger* of being long under the enemy's fire.

In all cases, it shall be the imperative duty of each captain to resort to every expedient for the protection of the motive and steering power of his vessel.

Vessels, in time of war, should only carry such

lights at night as may be absolutely necessary, and these should be carefully screened from outside view. When, under these instructions, the running lights are carried, covers should be placed over them, and hands stationed to remove and replace them as directed.

From dusk to daylight the bell and the pipe should be silent, and no loud noise of any kind allowed on board.

A man-of-war, in commission, should at all times be ready to cast loose her batteries, and throw open her magazines and shell-rooms at the first tap of the drum; and when anchored in any place where she may be assailed by the enemy, her engines should be "turned over" at least once in each watch, that she may be able to "go ahead" or "back" at a moment's notice. The signal of one's being attacked should be the discharge of a howitzer (that all friends within hearing may be notified) and the ringing of the ship's bell. At this signal the man at the slip-rope slips the cable, the engineer on watch closes his furnace doors, starts the engine (ahead or back as may be ordered), and the officers and men, without stopping to dress themselves, repair to their stations. A vessel thus drilled will never fall but with honor; and it should be a maxim with the navy that *none but a negligent officer will ever suffer himself to be surprised.*

No vessel in time of war should permit a *stranger* to approach her without being at quarters and *under way.*

When the fleet is engaged, if a vessel of it take fire so badly as to endanger her neighbors, her commanding officer shall at once leave his station and steer toward the enemy; and when it shall become apparent that he must abandon his vessel, he shall, before doing so, take care that she be pointed fair for the enemy's vessels, so that she may blow up, if possible, in their midst. While thus engaged, he should be careful to signal to his divisional commander, that a despatch boat may be sent to pick up or take off his officers and crew

In battle, no division, squadron, or vessel of the fleet shall separate from it to pursue any vessel or vessels of the enemy's fleet that may attempt to escape, unless signalled so to do by the admiral; but shall, on the contrary, do its utmost to assist in overwhelming and making certain the capture of such of the enemy as remain.

When the admiral's signals are obscured in action, it is the duty of his divisional officers to carry out energetically, and to the best of their judgment, his plan of battle (as previously made known to them); and the same may be said of the duty of each captain of a vessel with reference to his divisional or squadron commander; for it should be carefully borne in mind by each officer of the fleet that his duty requires him (when signals cannot be read) *to take advantage of any and every opportunity that may offer for damaging the enemy.*

Just before daylight, in time of war, every vessel, whether cruising singly or in company, is to be careful to see that her furnaces are in such a condition as shall enable her to put forth her utmost speed at the break of day, should circumstances require it.

Despatch vessels, in time of battle, may be advantageously employed, not only in carrying verbal and written orders, and in transmitting orders by signals to the various vessels of the fleet, but in putting prize crews on board captured vessels, in towing *into* action any vessels of the fleet whose steam power has become deranged or damaged, and in taking off, or picking up, the crews of sinking or burning vessels. They will, therefore, have their tow-lines ready and their boats clear for lowering at a moment's warning.

During an engagement, the officer commanding "the reserve" should, unless otherwise ordered, have his vessels well together, near the admiral's ship, on board of which he himself should remain (if circumstances will permit) until his services are required with his division, in order that he may be thoroughly posted as to the admiral's designs, and act accordingly when *the decisive moment has arrived.*

Vessels employed in reconnoitering should do so boldly, and not be satisfied until they have obtained a full knowledge of the strength of the enemy, of the course he is steering, of the disposition of his forces, etc., etc., which they should transmit with all possible speed, and in detail, through their divisional officers to the admiral.

Merchantmen under convoy should have a man-of-war as a leader, one on each flank, and one to bring up the rear.

The countersign should be signalled by the admiral every day at noon, and all vessels separating from the fleet should on their return to it, besides displaying their numbers, signal the countersign last given to them.

To demonstrate the weakness of the old order of battle in line ahead ("column of vessels"), now that steam has taken the place of sails, let us suppose that two fleets of equal strength fall in with each other; the one (A) formed in column of vessels, with its reserve (a) on the starboard—rear—flank, and steering W.; the other (B) in line, heading N., and bearing down to engage (A), with its reserve (b) posted in rear of its centre division.

Fig. 1.

Now there are two principal methods of attack open to B, viz.: to concentrate on the enemy's van or rear (Figs. 1 and 2), or to attempt to pierce his centre (Fig. 3). To attack the enemy's rear, let the assailing force before getting within range be thrown into echelon upon general signal 28 *(fleet N. W.);* then the enemy, seeing his van threatened, will immediately hurry his reserve up to its support. In the meantime B, who has simply intended his movement as a *feint,* is fleeting *his* reserve to the right,* and forming it into column of vessels; and so soon as his dispositions are made and he finds himself at the proper distance† from the enemy, he makes general signal 4 *(fleet N.E.);*‡ the enemy now discovering B's real objective will immediately reverse,§ during a part of which

* Should B's feint not induce A to transfer his reserve from rear to van, B should fleet *his* reserve to the left instead of to the right, and, going ahead at full speed, convert his feint into an attack of concentration on A's van.

† This will depend upon the speed at which the fleets are moving, etc., etc. The judgment of the admiral, guided by long experience, will not fail to tell him when the auspicious moment has arrived.

‡ A cannot decline a general engagement now, if he should desire to do so; for were he to attempt to make off, his stern would be exposed to the fire of his assailant.

§ Should A continue on his course, B's reserve should cross his stern and endeavor to come up on his starboard side, while the van and centre divisions, coming first N.W. and afterwards W. (supposing the strength of the vessels of those divisions to be in their broadside batteries), engage him on the port side, and the rear division, formed into column of vessels, endeavors to penetrate his column.

operation, he will be obliged to receive the whole broadside fire of B, with nothing but his bow guns to respond with ; and being next assailed vigorously by B, whose plan of attack has been deliberately conceived and determined upon, while he himself, in doubt and perplexity, is obliged to change his order of battle* at the very commencement of the conflict, A has nothing to expect but defeat.

FIG. 2.

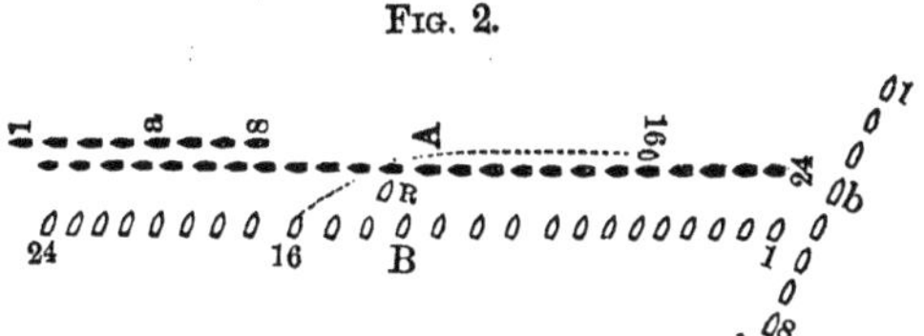

As the operation of concentrating on A's van would be similar to the above in every respect, except that in this case the *feint* would be towards

* A, being here attacked at the head (for, having reversed, his rear is now leading) by B's reserve, must inevitably be thrown into some confusion taking advantage of which, B's rear division, forming column of vessels from the right, should endeavor to break through A's column, as shown by the dotted line, and move up on the port side of his centre and van, already engaged on the starboard side by B's centre and van. (Fig. 2.)

his rear, we may now proceed to consider the attack on the centre. (Fig. 3.)

In this case B, closing up his right wing to within a half cable's length,* fleets his reserve to the left and moves down in line until at the proper distance from the enemy, when he throws his right wing and reserve into echelon by signaling 64 *(from the left, form echelon of vessels)*, leaving his left wing in line as at first; and when all his dispositions are made he signals both to the fleet and reserve, *N. W. full speed*, and sweeps to the attack as shown by the diagram.

Fig. 3.

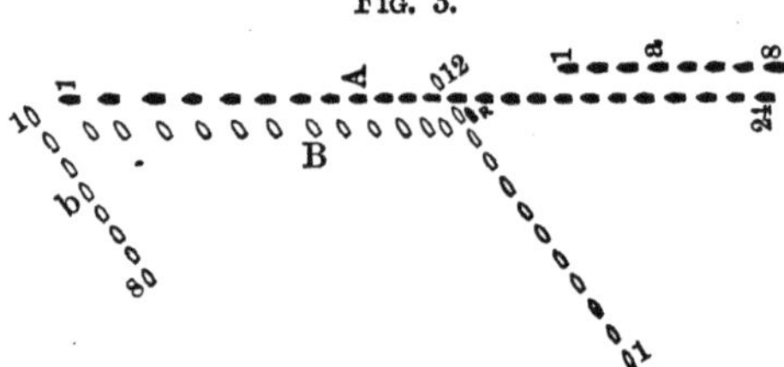

Here, as the right wing breaks through A's column, each vessel of it, manning both sides,

* Moving first in line, and afterward in echelon, he can do this in any ordinary weather without danger of collision. His object is, that his vessels may be well closed up when he signals N.W.

pours a raking broadside into his number twelve and number thirteen, and then following her next ahead comes sharp round the stern of number twelve, and moves up on the starboard side of A's centre and van, already engaged on the port side with B's left wing and attacked at the head by B's reserve.* I would remark here that the column designed to pierce the enemy's centre should always be led by a vessel of great artillery and *ramming* power, and should have abreast of it *(a little in advance)* the most formidable ram, or torpedo vessel, in the fleet, which should strike *at full speed* one of the enemy's vessels, thereby sinking or disabling it, and opening a passage *ahead* of the engaged vessel for the attacking column. The vessel marked R, in Figs. 2 and 3, is designed to represent this ram or torpedo vessel.

If the fleet were composed partly of vessels whose strength was in their broadside batteries, and in part of those whose fighting power was in their bows, the former, led as we have suggested, should form the column of penetration, while the latter, ranged in line, make the *direct* attack,

---

* This is a spirited, dashing attack, and properly conducted must succeed.

and the objective aimed at should always be, that wing of the enemy which appears to be composed of his weakest vessels.

Having explained above, wherein the order of battle in column of vessels is defective, it now remains to show how the attack of B Figs. 1 and 2) could have been frustrated by A, had the latter assumed an echelon formation. Let us suppose, then, A to be formed in column of vessels as before, and B bearing down in line to attack him. Now, when A observes that B is approaching, let him fleet his reserve abreast of his van division, and then signal to it and the rear division 78 (*forward into line, right oblique*), and afterwards 64 (*from the left, form echelon of vessels*); when the fleets will be in the position shown by Fig. 4.

FIG. 4.

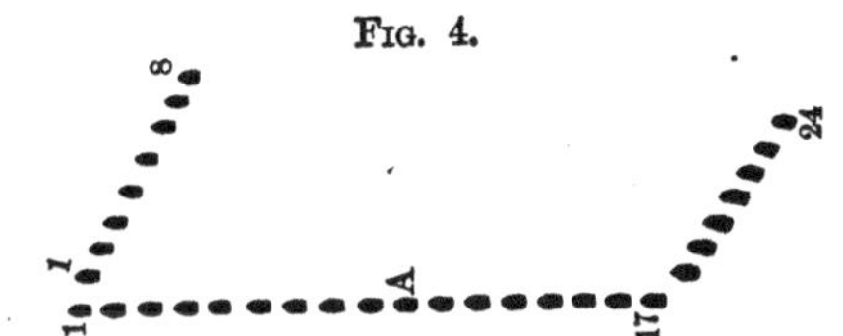

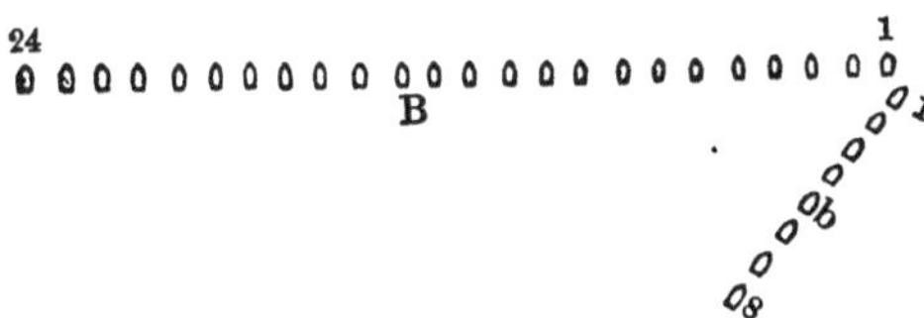

And so soon as B has approached so near that he (B) cannot decline an engagement, let A signal both to the fleet and reserve *S. W. full speed*, and attack at once, as shown by Fig. 5.

FIG. 5.

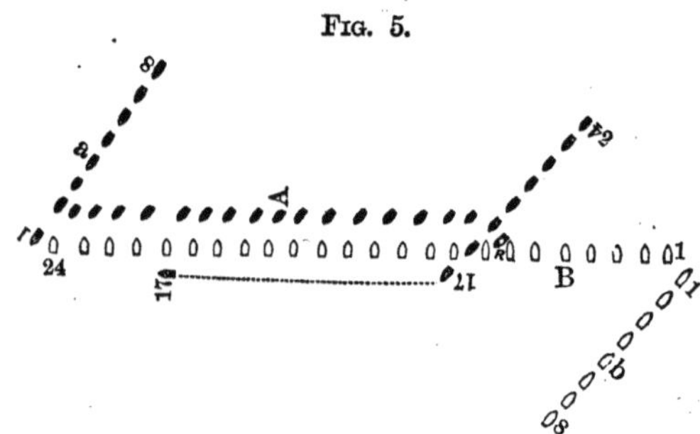

Here A's rear division breaks through the enemy's line, each vessel of it pouring a raking broadside into his number eight and number nine, and coming sharp round the stern of number nine, and then moving up astern of his centre and van, already engaged in front by A's centre and van, and assailed in flank by A's reserve.

In the second case (Fig. 3), where A finds B closing up his right wing, and forming it into echelon, with the intention of attacking his centre, he should make the same disposition of his vessels as before, except that it will be better to throw the whole left wing into echelon, thus:

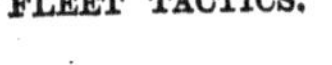
FIG. 6.

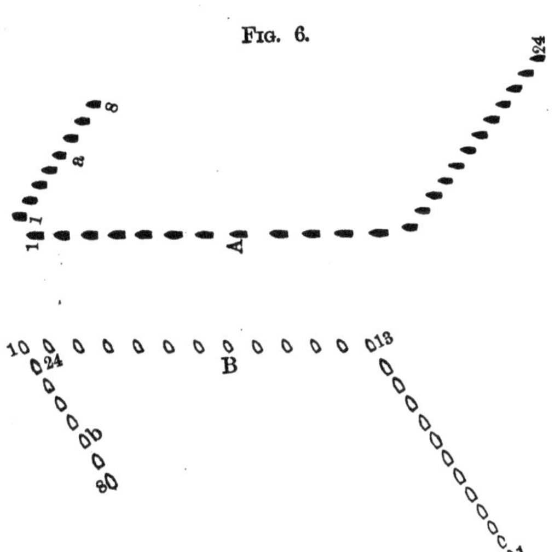

Now, let A preserve only such speed as shall enable B to come fairly down upon him, in the belief that he intends to engage in his present formation, until B has approached too near to retreat, when A makes signal *full speed*, and the instant he finds that he has *outflanked* B, he comes S.W. and signals to the right wing *slow*,* and attacks as shown by Fig. 7.

* His object in "slowing" his right, is that it may not interfere with the fire of his left wing and reserve in passing B's left wing.

FIG. 7.

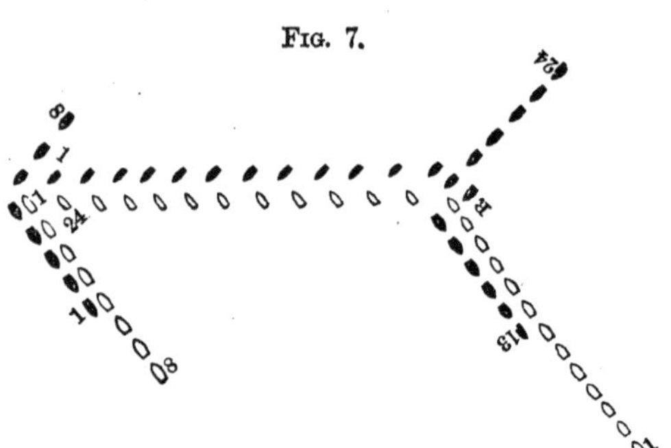

In this "cross attack" upon B's right wing and reserve,* it is manifest that the advantage would be altogother on the side of B's opponent, since the leading vessels of B's right wing would be exposed, not only to the raking broadside fire of such of A's vessels as were crossing their bows, but to the oblique fire of those that were coming up; the leaders of B's reserve being subject in

* When acting upon the defensive, it is, no doubt, prudent to keep a division strictly in reserve, somewhere near the centre of the fleet, in readiness to succor that portion of it which may be assailed; but, in assuming the offensive, I think the sooner the "reserve" is "put in" the better. Here it will be observed that both A and B have developed all their strength with the design of attacking; the former having gained the advantage solely by stratagem. A good military maxim to be deduced from our experience ashore and afloat during the rebellion would be, I should say: *On the offensive dare everything; on the defensive risk nothing.*

like manner to the direct raking fire of A's reserve, while his left wing, after being assailed on both flanks by the raking fire of A's left wing and reserve (to which would be added the oblique fire of A's reserve on its left), would be engaged in front by A's right wing.*

In conclusion, I would call attention to the necessity of guarding carefully the flanks of a steam fleet, since a neglect of this precaution, even were it composed of vessels with heavy broadside batteries, could not but be disastrous; while in a fleet of monitors, or rams, or of any class of vessels whose *weak points were abeam*, it would, most probably, lead to utter ruin; for the tactics of such a fleet would correspond almost exactly with those of an army, one of whose wings being "doubled up" entails confusion and ofttimes defeat upon the whole command.

---

* In all the above diagrams I have supposed A and B to have reached the place of combat, and, therefore, with their forces deployed in readiness to engage.

Before reaching the place of combat, however, the fleet should steam either in echelon of divisions, with each division formed into double echelon, from the centre, or into echelon of vessels from the right and left, or else in columns of divisions, with each division formed in double column on the centre; and a good tactician would be careful not to form it into order of battle until the time had come to engage, so that the enemy might be kept, up to the last moment, in ignorance of his designs.

The naval officer will, therefore, do well in the future to make a close study of military tactics, and especially of Jomini's "Art of War;" and he cannot fail to derive both pleasure and instruction from a perusal of Prescott's graphic description of the battle of Lepanto, where the Christian fleet, under Don Juan of Austria, broke forever the naval supremacy of the Turks.

INDEX TO SIGNALS

USED IN

MANŒUVRING BY THIS METHOD.

The author would suggest that the signal of manœuvre—to be hoisted at the fore—be the Union flag by day and a green lantern at night.

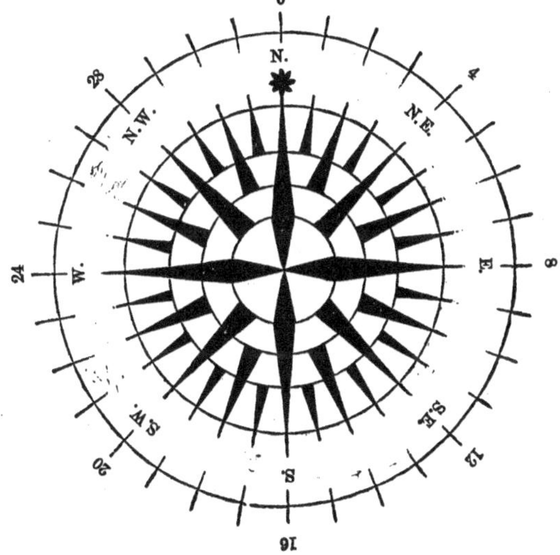

The signal will be dipped once and hoisted again for a quarter of a point, twice for half a point, and three times for three-quarters of a point—always counting to the right.

Column.

32. Fleet, division or squadron—from the right of—form column of vessels.—
33. Fleet, division or squadron—from the left of —form column of vessels.—
34. Fleet, division or squadron—form double column—right oblique.—
35. Fleet, division or squadron—form double column—left oblique.—
36. Fleet, division or squadron—from the right of—form double column.—
37. Fleet, division or squadron—from the left of —form double column.—
38. Fleet, division or squadron—form triple column—right oblique.—
39. Fleet, division or squadron—form triple column—left oblique.—
40. Fleet, division or squadron—from the right of—form triple column.—
41. Fleet, division or squadron—from the left of —form triple column.—

42. Fleet, division or squadron—form column of fours—right oblique.—
43. Fleet, division or squadron—form column of fours—left oblique.—
44. Fleet, division or squadron—from the right of—form column of fours.—
45. Fleet, division or squadron—from the left of—form column of fours—
46. Fleet or division—form column of squadrons —right oblique.—
47. Fleet or division—form column of squadrons —left oblique.
48. Fleet or division—from the right of—form column of squadrons.—
49. Fleet or division—from the left of—form column of squadrons.—
50. Fleet—form column of divisions—right oblique.
51. Fleet—form column of divisions—left oblique.—
52. Fleet—from the right of—form column of divisions.—
53. Fleet—from the left of—form column of divisions.
54. Fleet, division or squadron—from the centre of—form double column.—

55. Fleet or division—from the centre of—form column of squadrons.—
56. Fleet—from the centre of—form column of divisions.—
57. Fleet—on the division or squadron whose distinguishing pennant is shown above this signal—form columns of vessels abreast by divisions or squadrons—in natural order, and heading as shown by compass signal.—
58. Fleet—on the division or squadron whose distinguishing pennant is shown above this signal—form column of vessels abreast by divisions or squadrons—in order reversed, and heading as shown by compass signal.—
59. Fleet, division or squadron—form column of vessels—van in front.—
60. Fleet—on the division or squadron whose distinguishing pennant is shown above this signal—form column of vessels—van in front.
61. Fleet, division or squadron—form column of vessels—rear in front.—
62. Fleet—on the division or squadron whose distinguishing pennant is shown above this signal—form column of vessels—rear in front.—

Echelon (single).

63. Fleet, division or squadron—from the right of—form echelon of vessels.—
64. Fleet, division or squadron—from the left of —form echelon of vessels.—
65. Fleet or division—from the right of—form echelon of squadrons.—
66. Fleet or division—from the left of—form echelon of squadrons.—
67. Fleet—from the right of—form echelon of divisions.—
68. Fleet—from the left of—form echelon of divisions.—

Echelon (double).

69. Fleet, division or squadron—from the vessel whose distinguishing pennant is shown above this signal—form double echelon.—
70. Fleet, division or squadron—from the right and left of—on the vessel whose distinguishing pennant is shown above this signal—form double echelon.—
71. Fleet or division—from the squadron whose distinguishing pennant is shown above this signal—form double echelon of squadrons.—

72. Fleet or division—from the right and left squadrons of—on the squadron whose distinguishing pennant is shown above this signal—form double echelon of squadrons.—

73. Fleet, division or squadron—on the vessel whose distinguishing pennant is shown above this signal—form double echelon—right wing forward.—

74. Fleet, division or squadron—on the vessel whose distinguishing pennant is shown above this signal—form double echelon—left wing forward.—

75. Fleet, division or squadron—from the vessel whose distinguishing pennant is shown above this signal—form double echelon—right wing to the rear.—

76. Fleet, division or squadron—from the vessel whose distinguishing pennant is shown above this signal—form double echelon—left wing to the rear.—

Line.

77. Fleet, division or squadron—forward into line.—

78. Fleet, division or squadron—forward into line—right oblique.—

79. Fleet, division or squadron—forward into line—left oblique.—

Order.

80. Fleet, division or squadron—on the vessel, squadron or division whose distinguishing pennant is hoisted above this signal—form at half distance.—
81. Fleet, division or squadron—on the vessel, squadron or division whose distinguishing pennant is hoisted above this signal—form in close order.—
82. Fleet, division or squadron—on the vessel, squadron or division whose distinguishing pennant is hoisted above this signal—form in open order.—

Order of Battle.

83. Fleet, division or squadron—from the right of—in two lines—form order of battle.—
84. Fleet, division or squadron—from the left of —in two lines—form order of battle.—
85. Fleet, division or squadron—from the right of—in two columns—form order of battle.—
86. Fleet, division or squadron—from the left of —in two columns—form order of battle.

Speed.

87. Fleet, division, squadron or vessel—stop.—
88. Fleet, division, squadron or vessel—go ahead slow.—
89. Fleet, division, squadron or vessel—back.—
90. Fleet, division, squadron or vessel—steam four knots.—
91. Fleet, division, squadron or vessel—steam five knots.—
92. Fleet, division, squadron or vessel—steam six knots.—
93. Fleet, division, squadron or vessel—steam seven knots.—
94. Fleet, division, squadron or vessel—steam eight knots.—
95. Fleet, division, squadron or vessel—steam nine knots.—
96. Fleet, division, squadron or vessel—steam ten knots.—
97. Fleet, division, squadron or vessel—steam eleven knots.—
98. Fleet, division, squadron or vessel—steam twelve knots.
99. Fleet, division, squadron or vessel—steam at full speed.—

Steer.

100. Fleet, division or squadron—head of—steer as shown by compass signal.—
101. Fleet, division or squadron—starboard column of—steer as shown by compass signal.—
102. Fleet, division or squadron—port column of—steer as shown by compass signal.—
103. Fleet, division or squadron—head of starboard column of—steer as shown by compass signal.—
104. Fleet, division or squadron—head of port column of—steer as shown by compass signal.—
105. Fleet, division or squadron—right wing of—steer as shown by compass signal.—
106. Fleet, division or squadron—left wing of—steer as shown by compass signal.—

Wheel.

107. Fleet, division or squadron—by twos—wheel to the course indicated by compass signal.—
108. Fleet, division or squadron—by threes—wheel to the course indicated by compass signal.—

109. Fleet, division or squadron—by fours—wheel to the course indicated by compass signal.—
110. Fleet, or division—by squadrons—wheel to the course indicated by compass signal.—
111. Fleet—by divisions—wheel to the course indicated by compass signal.—
112. Fleet—wheel to the course indicated by compass signal.—

To show how signals are made by this method, let the reader refer to manœuvre 1, Fig. 16. Here the commander-in-chief hoists at the fore: first, the signal of manœuvre, next *general signal** (N.E.,) and finally, right vessel's distinguishing pennant, over No. 0 (N). At the same time he runs up at the main *general signal* 32 (From the right, form column of vessels).

The flag-ships of divisions repeat *all* these signals,† and so soon as the admiral has satisfied

* General signals apply to the fleet proper, and *not* to the *reserve*, which is not to regard any signal over which its distinguishing pennant is not hoisted.

† As (in this case) the vessel on the right of the fleet is not in either the centre or rear division, it may seem unnecessary for the flag-ships of those divisions to repeat the signals made to her by the admiral; but a good reason for

himself that they have been correctly repeated, he hauls them down, with the exception of the flag of manœuvre, which remains flying at his foremast head.

So soon as the signals of the divisional officers have been repeated* by the vessels of their respective divisions, they haul down No. 32 from the main, and hoist, in its stead, *general* signal 93 (steam seven knots); the flag-ship of the van division adding thereto right vessel's distinguishing pennant over 96 (steam ten knots); and when these last signals have been repeated, the divisional officers haul down everything but the flags of manœuvre. The admiral now understands that all are ready for the manœuvre; and so soon as he desires to execute it, he hauls down the flag of manœuvre; the divisional officers follow suit, and the *manœuvre* is commenced instantly, by the vessel on the extreme right of the fleet hoisting

---

their being required to repeat them is, that they may be often in a position where the admiral's signals can be more easily read by them, than by the flag-ship of the van division.

* Some officers may think it sufficient for these vessels to hoist here their answering pennants; but, to avoid mistakes, I think every signal of manœuvre should be repeated by the vessel or vessels to which it is made.

the guide flag* and keeping her course at the rate of ten knots an hour, while the rest of the fleet, putting their helms to port *simultaneously*, come N.E. and reduce their speed to seven knots an hour.

---

* At night a large lantern must take the place of the guide-flag.

Whenever any division, squadron or vessel of the fleet—other than the reserve—is intended to be released from obedience to a *general* signal, its distinguishing pennant must be hoisted *below* the signal.

*** *Any book in this Catalogue sent free by mail on receipt of price.*

VALUABLE

SCIENTIFIC BOOKS,

PUBLISHED BY

D. VAN NOSTRAND,

23 MURRAY STREET AND 27 WARREN STREET,

NEW YORK.

FRANCIS. Lowell Hydraulic Experiments, being a selection from Experiments on Hydraulic Motors, on the Flow of Water over Weirs, in Open Canals of Uniform Rectangular Section, and through submerged Orifices and diverging Tubes. Made at Lowell, Massachusetts. By James B. Francis, C. E. 2d edition, revised and enlarged, with many new experiments, and illustrated with twenty-three copperplate engravings. 1 vol. 4to, cloth......................$15 00

ROEBLING (J. A.) Long and Short Span Railway Bridges. By John A. Roebling, C. E. Illustrated with large copperplate engravings of plans and views. Imperial folio, cloth..... .... 25 00

CLARKE (T. C.) Description of the Iron Railway Bridge over the Mississippi River, at Quincy, Illinois. Thomas Curtis Clarke, Chief Engineer. Illustrated with 21 lithographed plans. 1 vol. 4to, cloth...... .... 7 50

TUNNER (P.) A Treatise on Roll-Turning for the Manufacture of Iron. By Peter Tunner. Translated and adapted by John B. Pearse, of the Pennylvania Steel Works, with numerous engravings wood cuts and folio atlas of plates.................$10 00

ISHERWOOD (B. F.) Engineering Precedents for Steam Machinery. Arranged in the most practical and useful manner for Engineers. By B. F. Isherwood, Civil Engineer, U. S. Navy. With Illustrations. Two volumes in one. 8vo, cloth........... $2 50

BAUERMAN. Treatise on the Metallurgy of Iron, containing outlines of the History of Iron Manufacture, methods of Assay, and analysis of Iron Ores, processes of manufacture of Iron and Steel, etc., etc. By H. Bauerman. First American edition. Revised and enlarged, with an Appendix on the Martin Process for making Steel, from the report of Abram S. Hewitt. Illustrated with numerous wood engravings. 12mo, cloth........................ .... 2 00

CAMPIN on the Construction of Iron Roofs. By Francis Campin. 8vo, with plates, cloth........... 3 00

COLLINS. The Private Book of Useful Alloys and Memoranda for Goldsmiths, Jewellers, &c. By James E. Collins. 18mo, cloth.................... 75

CIPHER AND SECRET LETTER AND TELEGRAPHIC CODE, with Hogg's Improvements. The most perfect secret code ever invented or discovered. Impossible to read without the key. By C. S. Larrabee. 18mo, cloth. 1 00

COLBURN. The Gas Works of London. By Zerah Colburn, C. E. 1 vol 12mo, boards.............. 60

CRAIG (B. F.) Weights and Measures. An account of the Decimal System, with Tables of Conversion for Commercial and Scientific Uses. By B. F. Craig, M.D. 1 vol. square 32mo, limp cloth.... 50

NUGENT. Treatise on Optics; or, Light and Sight, theoretically and practically treated; with the application to Fine Art and Industrial Pursuits. By E. Nugent. With one hundred and three illustrations. 12mo, cloth.. 2 00

GLYNN (J.) Treatise on the Power of Water, as applied to drive Flour Mills, and to give motion to Turbines and other Hydrostatic Engines. By Joseph

Glynn. Third edition, revised and enlarged, with numerous illustrations. 12mo, cloth. $1 00

HUMBER. A Handy Book for the Calculation of Strains in Girders and similar Structures, and their Strength, consisting of Formulæ and corresponding Diagrams, with numerous details for practical application. By William Humber. 12mo, fully illustrated, cloth.................................. 2 50

GRUNER. The Manufacture of Steel. By M. L. Gruner. Translated from the French, by Lenox Smith, with an appendix on the Bessamer process in the United States, by the translator. Illustrated by Lithographed drawings and wood cuts. 8vo, cloth.. 3 50

AUCHINCLOSS. Link and Valve Motions Simplified. Illustrated with 37 wood-cuts, and 21 lithographic plates, together with a Travel Scale, and numerous useful Tables. By W. S. Auchincloss. 8vo, cloth.. 3 00

VAN BUREN. Investigations of Formulas, for the strength of the Iron parts of Steam Machinery. By J. D. Van Buren, Jr., C. E. Illustrated, 8vo, cloth. 2 00

JOYNSON. Designing and Construction of Machine Gearing. Illustrated, 8vo, cloth.................. 2 00

GILLMORE. Coignet Beton and other Artificial Stone. By Q. A. Gillmore, Major U. S. Corps Engineers. 9 plates, views, &c. 8vo, cloth.................... 2 50

SAELTZER. Treattse on Acoustics in connection with Ventilation. By Alexander Saeltzer, Architect. 12mo, cloth.................................. 2 00

THE EARTH'S CRUST. A handy Outline of Geology. By David Page. Illustrated, 18mo, cloth.... 75

DICTIONARY of Manufactures, Mining, Machinery, and the Industrial Arts. By George Dodd. 12mo, cloth.. 2 00

FRANCIS. On the Strength of Cast-Iron Pillars, with Tables for the use of Engineers, Architects, and Builders. By James B. Francis, Civil Engineer. 1 vol. 8vo, cloth.................................. 2 00

GILLMORE (Gen. Q. A.) Treatise on Limes, Hydraulic Cements, and Mortars. Papers on Practical Engineering, U. S. Engineer Department, No. 9, containing Reports of numerous Experiments conducted in New York City, during the years 1858 to 1861, inclusive. By Q. A. Gillmore, Bvt. Maj-Gen., U. S. A., Major, Corps of Engineers. With numerous illustrations. 1 vol, 8vo, cloth.............. $4 00

HARRISON. The Mechanic's Tool Book, with Practical Rules and Suggestions for Use of Machinists, Iron Workers, and others. By W. B. Harrison, associate editor of the "American Artisan." Illustrated with 44 engravings. 12mo, cloth............ 1 50

HENRICI (Olaus). Skeleton Structures, especially in their application to the Building of Steel and Iron Bridges. By Olaus Henrici. With folding plates and diagrams. 1 vol. 8vo, cloth..... .... 3 00

HEWSON (Wm.) Principles and Practice of Embanking Lands from River Floods, as applied to the Levees of the Mississippi. By William Hewson, Civil Engineer. 1 vol. 8vo, cloth....................... 2 00

HOLLEY (A. L.) Railway Practice. American and European Railway Practice, in the economical Generation of Steam, including the Materials and Construction of Coal-burning Boilers, Combustion, the Variable Blast, Vaporization, Circulation, Superheating, Supplying and Heating Feed-water, etc., and the Adaptation of Wood and Coke-burning Engines to Coal-burning; and in Permanent Way, including Road-bed, Sleepers, Rails, Joint-fastenings, Street Railways, etc., etc. By Alexander L. Holley, B. P. With 77 lithographed plates. 1 vol. folio, cloth.... 12 00

KING (W. H.) Lessons and Practical Notes on Steam, the Steam Engine, Propellers, etc., etc., for Young Marine Engineers, Students, and others. By the late W. H. King, U. S. Navy. Revised by Chief Engineer J. W. King, U. S. Navy. Twelfth edition, enlarged. 8vo, cloth. 2 00

MINIFIE (Wm.) Mechanical Drawing. A Text-Book of Geometrical Drawing for the use of Mechanics and Schools, in which the Definitions and Rules of Geometry are familiarly explained; the Practical Problems are arranged, from the most simple to the more complex, and in their description technicalities are avoided as much as possible. With illustrations for Drawing Plans, Sections, and Elevations of Railways and Machinery; an Introduction to Isometrical Drawing, and an Essay on Linear Perspective and Shadows. Illustrated with over 200 diagrams engraved on steel. By Wm. Minifie, Architect. Seventh edition. With an Appendix on the Theory and Application of Colors. 1 vol. 8vo, cloth.......... $4 00

"It is the best work on Drawing that we have ever seen, and is especially a text-book of Geometrical Drawing for the use of Mechanics and Schools. No young Mechanic, such as a Machinist, Engineer, Cabinet-maker, Millwright, or Carpenter, should be without it."—*Scientific American.*

—— Geometrical Drawing. Abridged from the octavo edition, for the use of Schools. Illustrated with 48 steel plates. Fifth edition. 1 vol. 12mo, cloth.... 2 00

STILLMAN (Paul.) Steam Engine Indicator, and the Improved Manometer Steam and Vacuum Gauges—their Utility and Application. By Paul Stillman. New edition. 1 vol. 12mo, flexible cloth.......... 1 00

SWEET (S. H.) Special Report on Coal; showing its Distribution, Classification, and cost delivered over different routes to various points in the State of New York, and the principal cities on the Atlantic Coast. By S. H. Sweet. With maps, 1 vol. 8vo, cloth..... 3 00

WALKER (W. H.) Screw Propulsion. Notes on Screw Propulsion: its Rise and History. By Capt. W. H. Walker, U. S. Navy. 1 vol. 8vo, cloth..... 75

WARD (J. H.) Steam for the Million. A popular Treatise on Steam and its Application to the Useful Arts, especially to Navigation. By J. H. Ward, Commander U. S. Navy. New and revised edition. 1 vol. 8vo, cloth.............................. 1 00

WEISBACH (Julius). Principles of the Mechanics of Machinery and Engineering. By Dr. Julius Weisbach, of Freiburg. Translated from the last German edition. Vol. I., 8vo, cloth.. .... 10 00

DIEDRICH. The Theory of Strains, a Compendium for the calculation and construction of Bridges, Roofs, and Cranes, with the application of Trigonometrical Notes, containing the most comprehensive information in regard to the Resulting strains for a permanent Load, as also for a combined (Permanent and Rolling) Load. In two sections, adadted to the requirements of the present time. By John Diedrich, C. E. Illustrated by numerous plates and diagrams. 8vo, cloth.................................. 5 00

WILLIAMSON (R. S.) On the use of the Barometer on Surveys and Reconnoissances. Part I. Meteorology in its Connection with Hypsometry. Part II. Barometric Hypsometry. By R. S. Wiliamson, Bvt. Lieut.-Col. U. S. A., Major Corps of Engineers. With Illustrative Tables and Engravings. Paper No. 15, Professional Papers, Corps of Engineers. 1 vol. 4to, cloth... .... 15 00

POOK (S. M.) Method of Comparing the Lines and Draughting Vessels Propelled by Sail or Steam. Including a chapter on Laying off on the Mould-Loft Floor. By Samuel M. Pook, Naval Constructor. 1 vol. 8vo, with illustrations, cloth............ 5 00

ALEXANDER (J. H.) Universal Dictionary of Weights and Measures, Ancient and Modern, reduced to the standards of the United States of America. By J. H. Alexander. New edition, enlarged. 1 vol. 8vo, cloth.................................. 3 50

BROOKLYN WATER WORKS. Containing a Descriptive Account of the Construction of the Works, and also Reports on the Brooklyn, Hartford, Belleville and Cambridge Pumping Engines. With illustrations. 1 vol. folio, cloth..........................

RICHARDS' INDICATOR. A Treatise on the Richards Steam Engine Indicator, with an Appendix by F. W. Bacon, M. E. 18mo, flexible, cloth.......... 1 00

POPE Modern Practice of the Electric Telegraph. A Hand Book for Electricians and operators. By Frank L. Pope. Eighth edition, revised and enlarged, and fully illlustrated. 8vo, cloth........................ $2.00

"There is no other work of this kind in the English language that contains in so small a compass so much practical information in the application of galvanic electricity to telegraphy. It should be in the hands of every one interested in telegraphy, or the use of Batteries for other purposes."

MORSE. Examination of the Telegraphic Apparatus and the Processes in Telegraphy. By Samuel F. Morse, LL.D., U. S. Commissioner Paris Universal Exposition, 1867. Illustrated, 8vo, cloth.......... $2 00

SABINE. History and Progress of the Electric Telegraph, with descriptions of some of the apparatus. By Robert Sabine, C. E. Second edition, with additions, 12mo, cloth...... .... 1 25

CULLEY. A Hand-Book of Practical Telegraphy. By R. S. Culley, Engineer to the Electric and International Telegraph Company. Fourth edition, revised and enlarged. Illustrated 8vo, cloth.............. 5 00

BENET. Electro-Ballistic Machines, and the Schultz Chronoscope. By Lieut.-Col. S. V. Benet, Captain of Ordnance, U. S. Army. Illustrated. second edition, 4to, cloth.................................. 3 00

MICHAELIS. The Le Boulenge Chronograph, with three Lithograph folding plates of illustrations. By Brevet Captain O. E. Michaelis, First Lieutenant Ordnance Corps, U. S. Army, 4to, cloth.......... 3 00

ENGINEERING FACTS AND FIGURES An Annual Register of Progress in Mechanical Engineering and Construction. for the years 1863, 64, 65, 66, 67, 68. Fully illustrated, 6 vols. 18mo, cloth, $2.50 per vol., each volume sold separately..............

HAMILTON. Useful Information for Railway Men. Compiled by W. G. Hamilton, Engineer. Fifth edition, revised and enlarged, 562 pages Pocket form. Morocco, gilt.................................... 2 00

STUART. The Civil and Military Engineers of America. By Gen. C. B. Stuart. With 9 finely executed portraits of eminent engineers, and illustrated by engravings of some of the most important works constructed in America. 8vo, cloth.................. $5 00

STONEY. The Theory of Strains in Girders and similar structures, with observations on the application of Theory to Practice, and Tables of Strength and other properties of Materials. By Bindon B. Stoney, B. A. New and revised edition, enlarged, with numerous engravings on wood, by Oldham. Royal 8vo, 664 pages. Complete in one volume. 8vo, cloth. 15 00

SHREVE. A Treatise on the Strength of Bridges and Roofs. Comprising the determination of Algebraic formulas for strains in Horizontal, Inclined or Rafter, Triangular, Bowstring, Lenticular and other Trusses, from fixed and moving loads, with practical applications and examples, for the use of Students and Engineers. By Samuel H. Shreve, A. M., Civil Engineer. 87 wood cut illustrations. 8vo, cloth............... 5 00

MERRILL. Iron Truss Bridges for Railroads. The method of calculating strains in Trusses, with a careful comparison of the most prominent Trusses, in reference to economy in combination, etc., etc. By Brevet. Col. William E. Merrill, U S. A., Major Corps of Engineers, with nine lithographed plates of Illustrations. 4to, cloth......................... 5 00

WHIPPLE. An Elementary and Practical Treatise on Bridge Building. An enlarged and improved edition of the author's original work. By S. Whipple, C. E., inventor of the Whipple Bridges, &c. Illustrated 8vo, cloth.................................... 4 00

THE KANSAS CITY BRIDGE. With an account of the Regimen of the Missouri River, and a description of the methods used for Founding in that River. By O Chanute, Chief Engineer, and George Morrison, Assistant Engineer. Illustrated with five lithographic views and twelve plates of plans. 4to, cloth, 6 00

MAC CORD. A Practical Treatise on the Slide Valve by Eccentrics, examining by methods the action of the Eccentric upon the Slide Valve, and explaining the Practical processes of laying out the movements, adapting the valve for its various duties in the steam engine. For the use of Engineers, Draughtsmen, Machinists, and Students of Valve Motions in general. By C. W. Mac Cord, A. M., Professor of Mechanical Drawing, Stevens' Institute of Technology, Hoboken, N. J. Illustrated by 8 full page copperplates. 4to. cloth $4 00

KIRKWOOD. Report on the Filtration of River Waters, for the supply of cities, as practised in Europe, made to the Board of Water Commissioners of the City of St. Louis. By James P. Kirkwood. Illustrated by 30 double plate engravings. 4to, cloth, 15 00

PLATTNER. Manual of Qualitative and Quantitative Analysis with the Blow Pipe. From the last German edition, revised and enlarged. By Prof. Th. Richter, of the Royal Saxon Mining Academy. Translated by Prof. H. B. Cornwall, Assistant in the Columbia School of Mines, New York, assisted by John H. Caswell. Illustrated with 87 wood cuts, and one lithographic plate. Second edition, revised, 560 pages, 8vo, cloth 7 50

PLYMPTON. The Blow Pipe. A system of Instruction in its practical use being a graduated course of analysis for the use of students, and all those engaged in the examination of metallic combinations. Second edition, with an appendix and a copious index. By Prof. Geo. W. Plympton, of the Polytechnic Institute, Brooklyn, N. Y. 12mo, cloth 2 00

PYNCHON. Introduction to Chemical Physics, designed for the use of Academies, Colleges and High Schools. Illustrated with numerous engravings, and containing copious experiments with directions for preparing them. By Thomas Ruggles Pynchon, M. A., Professor of Chemistry and the Natural Sciences, Trinity College, Hartford. New edition, revised and enlarged and illustrated by 269 illustrations on wood. Crown, 8vo. cloth 3 00

ELIOT AND STORER. A compendious Manual of Qualitative Chemical Analysis. By Charles W. Eliot and Frank H. Storer. Revised with the Co-operation of the authors. By William R. Nichols, Professor of Chemistry in the Massachusetts Institute of Technology Illustrated, 12mo, cloth....... $1 50

RAMMELSBERG. Guide to a course of Quantitative Chemical Analysis, especially of Minerals and Furnace Products. Illustrated by Examples By C. F. Rammelsberg. Translated by J. Towler, M. D. 8vo, cloth.. 2 25

EGLESTON. Lectures on Descriptive Mineralogy, delivered at the School of Mines, Columbia College. By Professor T. Egleston. Illustrated by 34 Lithographic Plates. 8vo, cloth........................ 4 50

MITCHELL. A Manual of Practical Assaying. By John Mitchell. Third edition. Edited by William Crookes, F. R. S. 8vo, cloth.... .... 10 00

WATT'S Dictionary of Chemistry. New and Revised edition complete in 6 vols 8vo cloth, $62.00 Supplementary volume sold separately. Price, cloth... 9 00

RANDALL. Quartz Operators Hand-Book. By P. M. Randall. New edition, revised and enlarged, fully illustrated. 12mo, cloth 2 00

SILVERSMITH. A Practical Hand-Book for Miners, Metallurgists, and Assayers, comprising the most recent improvements in the disintegration amalgamation, smelting, and parting of the precious ores, with a comprehensive Digest of the Mining Laws Greatly augmented, revised and corrected. By Julius Silversmith Fourth edition. Profusely illustrated. 12mo, cloth... 3 00

THE USEFUL METALS AND THEIR ALLOYS, including Mining Ventilation, Mining Jurisprudence, and Metallurgic Chemistry employed in the conversion of Iron, Copper, Tin, Zinc, Antimony and Lead ores, with their applications to the Industrial Arts. By Scoffren, Truan, Clay, Oxland, Fairbairn, and others. Fifth edition, half calf..................... 3 75

JOYNSON. The Metals used in construction, Iron, Steel, Bessemer Metal, etc., etc. By F. H. Joynson. Illustrated, 12mo, cloth.............................. $0 75

VON COTTA. Treatise on Ore Deposits. By Bernhard Von Cotta, Professor of Geology in the Royal School of Mines, Freidberg, Saxony. Translated from the second German edition, by Frederick Prime, Jr., Mining Engineer, and revised by the author, with numerous illustrations. 8vo, cloth....... 4 00

URE. Dictionary of Arts, Manufactures and Mines By Andrew Ure, M.D. Sixth edition, edited by Robert Hunt, F. R. S, greatly enlarged and re-written. London, 1872. 3 vols 8vo, cloth, $25.00. Half Russia.................................... 37 50

BELL. Chemical Phenomena of Iron Smelting. An experimental and practical examination of the circumstances which determine the capacity of the Blast Furnace, The Temperature of the air, and the proper condition of the Materials to be operated upon. By I. Lowthian Bell. 8vo, cloth............ 6 00

ROGERS. The Geology of Pennsylvania. A Government survey, with a general view of the Geology of the United States, Essays on the Coal Formation and its Fossils, and a description of the Coal Fields of North America and Great Britain. By Henry Darwin Rogers, late State Geologist of Pennsylvania, Splendidly illustrated with Plates and Engravings in the text. 3 vols, 4to, cloth with Portfolio of Maps. 30 00

BURGH. Modern Marine Engineering, applied to Paddle and Screw Propulsion. Consisting of 36 colored plates, 259 Practical Wood Cut Illustrations, and 403 pages of descriptive matter, the whole being an exposition of the present practice of James Watt & Co., J. & G. Rennie, R. Napier & Sons, and other celebrated firms, by N. P. Burgh, Engineer, thick 4to, vol., cloth, $25.00; half mor.. ..... 30 00

BARTOL. Treatise on the Marine Boilers of the United States. By B. H Bartol. Illustrated, 8vo, cloth... 1 50

BOURNE. Treatise on the Steam Engine in its various applications to Mines, Mills, Steam Navigation, Railways, and Agriculture, with the theoretical investigations respecting the Motive Power of Heat, and the proper proportions of steam engines. Elaborate tables of the right dimensions of every part, and Practical Instructions for the manufacture and management of every species of Engine in actual use. By John Bourne, being the ninth edition of "A Treatise on the Steam Engine," by the "Artizan Club." Illustrated by 38 plates and 546 wood cuts. 4to, cloth.............................. ....$15 00

STUART. The Naval Dry Docks of the United States. By Charles B. Stuart late Engineer-in-Chief of the U. S. Navy. Illustrated with 24 engravings on steel. Fourth edition, cloth.................... 6 00

EADS. System of Naval Defences. By James B. Eads, C. E., with 10 illustrations, 4to, cloth........ 5 00

FOSTER. Submarine Blasting in Boston Harbor, Massachusetts. Removal of Tower and Corwin Rocks. By J. G. Foster, Lieut.-Col. of Engineers, U. S. Army. Illustrated with seven plates, 4to, cloth.. 3 50

BARNES Submarine Warfare, offensive and defensive, including a discussion of the offensive Torpedo System, its effects upon Iron Clad Ship Systems and influence upon future naval wars. By Lieut.-Commander J. S. Barnes, U. S. N., with twenty lithographic plates and many wood cuts. 8vo, cloth... . 5 00

HOLLEY. A Treatise on Ordnance and Armor, embracing descriptions, discussions, and professional opinions concerning the materials, fabrication, requirements, capabilities, and endurance of European and American Guns, for Naval, Sea Coast, and Iron Clad Warfare, and their Rifling, Projectiles, and Breech-Loading; also, results of experiments against armor, from official records, with an appendix referring to Gun Cotton, Hooped Guns, etc., etc. By Alexander L. Holley, B. P., 948 pages, 493 engravings, and 147 Tables of Results, etc, 8vo, half roan. 10 00

SIMMS. A Treatise on the Principles and Practice of Levelling, showing its application to purposes of Railway Engineering and the Construction of Roads, &c. By Frederick W. Simms, C. E. From the 5th London edition, revised and corrected, with the addition of Mr. Laws's Practical Examples for setting out Railway Curves. Illustrated with three Lithographic plates and numerous wood cuts. 8vo, cloth. $2 50

BURT. Key to the Solar Compass, and Surveyor's Companion; comprising all the rules necessary for use in the field; also description of the Linear Surveys and Public Land System of the United States, Notes on the Barometer, suggestions for an outfit for a survey of four months, etc By W. A. Burt, U. S. Deputy Surveyor. Second edition. Pocket book form, tuck.. 2 50

THE PLANE TABLE. Its uses in Topographical Surveying, from the Papers of the U. S. Coast Survey. Illustrated, 8vo, cloth........................ 2 00

"This work gives a description of the Plane Table, employed at the U. S. Coast Survey office, and the manner of using it."

JEFFER'S. Nautical Surveying. By W. N. Jeffers, Captain U. S. Navy. Illustrated with 9 copperplates and 31 wood cut illustrations. 8vo, cloth........... 5 00

CHAUVENET. New method of correcting Lunar Distances, and improved method of Finding the error and rate of a chronometer, by equal altitudes. By W. Chauvenet, LL D. 8vo, cloth................. 2 00

BRUNNOW. Spherical Astronomy. By F. Brunnow, Ph. Dr. Translated by the author from the second German edition. 8vo, cloth...................... 6 50

PEIRCE. System of Analytic Mechanics. By Benjamin Peirce. 4to, cloth.......................... 10 00

COFFIN. Navigation and Nautical Astronomy. Prepared for the use of the U. S. Naval Academy. By Prof. J H. C. Coffin. Fifth edition. 52 wood cut illustrations. 12mo, cloth.......................... 3 50

CLARK. Theoretical Navigation and Nautical Astronomy. By Lieut. Lewis Clark, U. S. N. Illustrated with 41 wood cuts. 8vo, cloth.................... $3 00

HASKINS. The Galvanometer and its Uses. A Manual for Electricians and Students. By C. H. Haskins. 12mo, pocket form, morocco. (In press).....

GOUGE. New System of Ventilation, which has been thoroughly tested, under the patronage of many distinguished persons. By Henry A. Gouge. With many illustrations. 8vo, cloth.................... 2 00

BECKWITH. Observations on the Materials and Manufacture of Terra-Cotta, Stone Ware, Fire Brick, Porcelain and Encaustic Tiles, with remarks on the products exhibited at the London International Exhibition, 1871. By Arthur Beckwith, C. E. 8vo, paper.. 60

MORFIT. A Practical Treatise on Pure Fertilizers, and the chemical conversion of Rock Guano, Marlstones, Coprolites, and the Crude Phosphates of Lime and Alumina generally, into various valuable products. By Campbell Morfit, M.D., with 28 illustrative plates, 8vo, cloth.................................... 20 00

BARNARD. The Metric System of Weights and Measures. An address delivered before the convocation of the University of the State of New York, at Albany, August, 1871. By F. A. P. Barnard, LL.D., President of Columbia College, New York. Second edition from the revised edition, printed for the Trustees of Columbia College. Tinted paper, 8vo, cloth 3 00

—— Report on Machinery and Processes on the Industrial Arts and Apparatus of the Exact Sciences. By F. A. P. Barnard, LL.D. Paris Universal Exposition, 1867. Illustrated, 8vo, cloth.............. 5 00

BARLOW. Tables of Squares, Cubes, Square Roots, Cube Roots, Reciprocals of all integer numbers up to 10,000. New edition, 12mo, cloth................. 2 50

MYER. Manual of Signals, for the use of Signal officers in the Field, and for Military and Naval Students, Military Schools, etc. A new edition enlarged and illustrated By Brig. General Albert J. Myer, Chief Signal Officer of the army, Colonel of the Signal Corps during the War of the Rebellion. 12mo, 48 plates, full Roan.. $5 00

WILLIAMSON. Practical Tables in Meteorology and Hypsometry, in connection with the use of the Barometer. By Col. R. S. Williamson, U. S. A. 4to, cloth.. 2 50

THE YOUNG MECHANIC. Containing directions for the use of all kinds of tools, and for the construction of Steam Engines and Mechanical Models, including the Art of Turning in Wood and Metal. By the author "The Lathe and its Uses," etc. From the English edition with corrections. Illustrated, 12mo, cloth.. 1 75

PICKERT AND METCALF. The Art of Graining. How Acquired and How Produced, with description of colors, and their application. By Charles Pickert and Abraham Metcalf. Beautifully illustrated with 42 tinted plates of the various woods used in interior finishing. Tinted paper, 4to, cloth................ 10 00

HUNT. Designs for the Gateways of the Southern Entrances to the Central Park. By Richard M. Hunt. With a description of the designs. 4to. cloth...... 5 00

LAZELLE. One Law in Nature. By Capt. H. M. Lazelle, U. S. A. A new Corpuscular Theory, comprehending Unity of Force, Identity of Matter, and its Multiple Atom Constitution, applied to the Physical Affections or Modes of Energy. 12mo, cloth... 1 50

PETERS Notes on the Origin, Nature, Prevention, and Treatment of Asiatic Cholera. By John C. Peters, M. D. Second edition, with an Appendix. 12mo, cloth.. 1 50

BOYNTON. History of West Point, its Military Importance during the American Revolution, and the Origin and History of the U. S. Military Academy. By Bvt. Major C. E. Boynton, A.M., Adjutant of the Military Academy. Second edition, 416 pp. 8vo, printed on tinted paper, beautifully illustrated with 36 maps and fine engravings, chiefly from photographs taken on the spot by the author. Extra cloth.... $3 50

WOOD. West Point Scrap Book, being a collection of Legends, Stories, Songs, etc., of the U S Military Academy. By Lieut. O. E. Wood, U. S. A. Illustrated by 69 engravings and a copperplate map. Beautifully printed on tinted paper. 8vo, cloth..... 5 00

WEST POINT LIFE. A Poem read before the Dialectic Society of the United States Military Academy. Illustrated with Pen-and-Ink Sketches. By a Cadet. To which is added the song, "Benny Havens, oh!" oblong 8vo, 21 full page illustrations, cloth......... 2 50

GUIDE TO WEST POINT and the U. S. Military Academy, with maps and engravings, 18mo, blue cloth, flexible.... 1 00

HENRY Military Record of Civilian Appointments in the United States Army By Guy V. Henry, Brevet Colonel and Captain First United States Artillery, Late Colonel and Brevet Brigadier General, United States Volunteers Vol. 1 now ready. Vol. 2 in press. 8vo, per volume, cloth.... 5 00

HAMERSLY Records of Living Officers of the U. S. Navy and Marine Corps. Compiled from official sources. By Lewis R. Hamersly, late Lieutenant U. S. Marine Corps. Revised edition, 8vo, cloth... 5 00

MOORE. Portrait Gallery of the War. Civil, Military and Naval. A Biographical record, edited by Frank Moore. 60 fine portraits on steel. Royal 8vo, cloth.... 6 00